RECLAMATION AND REPROCESSING
OF SPENT SOLVENTS

RECLAMATION AND REPROCESSING OF SPENT SOLVENTS

by

Arthur R. Tarrer

Department of Chemical Engineering
Auburn University
Auburn, Alabama

Bernard A. Donahue, Seshasayi Dharmavaram

U.S. Army Construction Engineering Research Laboratory
Champaign, Illinois

Surendra B. Joshi

U.S. Air Force Engineering and Services Center
Tyndall Air Force Base, Florida

NOYES DATA CORPORATION
Park Ridge, New Jersey, U.S.A.

Copyright © 1989 by Noyes Data Corporation
Library of Congress Catalog Card Number: 89-22887
ISBN: 0-8155-1222-8
ISSN: 0090-516X
Printed in the United States

Published in the United States of America by
Noyes Data Corporation
Mill Road, Park Ridge, New Jersey 07656

10 9 8 7 6 5 4 3 2 1

Library of Congress Cataloging-in-Publication Data

Reclamation and reprocessing of spent solvents / by Arthur R. Tarrer
    ... [et al.].
            p.   cm. -- (Pollution technology review, ISSN 0090-516X ; no.
    175)
      Includes bibliographical references.
      ISBN 0-8155-1222-8 :
      1. Solvents--Recycling. I. Tarrer, Arthur R. II. Series.
    TP247.5.R43   1989
    661'.807--dc20   89-22887
                                                      CIP

Foreword

This book describes reclamation and reprocessing of spent cleaning solvents, as well as tests to determine spent solvent quality. Large amounts of waste solvents are generated each year, and standard practice has previously been to discard these solvents. However, increasingly stringent environmental regulations, coupled with the soaring costs of waste disposal and new solvent purchase, have led to the search for safe, cost-effective methods of managing waste solvents. This study was prepared for the U.S. Army, but will be applicable to industrial users as well.

Reclamation is feasible because solvents usually do not break down chemically during cleaning operations for which they are used. However, this type of recycling has seen limited success in the past due to the lack of scientific tests and criteria for judging a solvent as spent.

Part I of the book describes the evaluation of tests for assessing the quality of used cold-cleaning solvents. From the findings, guidelines have been generated that will allow managers and those responsible for environmental compliance to quickly assess a used solvent's quality and determine if it should be reused or recycled. Reclamation techniques, their advantages and drawbacks, are also presented.

Part II covers tests for quality evaluation of used halogenated solvents employed primarily in cleaning. In addition to testing for solvent quality, chlorinated solvents must be checked for inhibitor concentration because adequate inhibitor levels are critical to safe, effective use of halogenated solvents. Inhibitors are identified and their reactions studied, as well as their usage over time. Finally, reclamation methods were assessed to determine recycling feasibility.

The information in the book is from the following documents:

Used Solvent Testing and Reclamation, Volume I: Cold-Cleaning Solvents prepared by Arthur R. Tarrer, Auburn University; Bernard A. Donahue and Seshasayi Dharmavaram, U.S. Army Construction Engineering Research Laboratory; and Surendra B. Joshi, U.S. Air Force Engineering and Services Center; for the U.S. Army Engineering and Housing Support Center, December 1988.

Used Solvent Testing and Reclamation, Volume II: Vapor Degreasing and Precision Cleaning Solvents, prepared by Arthur R. Tarrer, Auburn University; Bernard A. Donahue and Seshasayi Dharmavaram, U.S. Army Construction Engineering Research Laboratory; and Surendra B. Joshi, U.S. Air Force Engineering and Housing Support Center, December 1988.

The table of contents is organized in such a way as to serve as a subject index and provides easy access to the information contained in the book.

Advanced composition and production methods developed by Noyes Data Corporation are employed to bring this durably bound book to you in a minimum of time. Special techniques are used to close the gap between "manuscript" and "completed book." In order to keep the price of the book to a reasonable level, it has been partially reproduced by photo-offset directly from the original reports and the cost saving passed on to the reader. Due to this method of publishing, certain portions of the book may be less legible than desired.

ACKNOWLEDGMENTS

This report summarizes work done by Auburn University, Department of Chemical Engineering. The work was performed under the direction of Dr. A.R. Tarrer. The contributions of Dr. Mahmud A. Rahman, David L. Phillips, Dr. James A. Guin, Dr. Christine W. Curtis, and Bill L. Brady, Jr. of Auburn University in completing this report are acknowledged. The USA-CERL Principal Investigator was Bernard Donahue; Seshasayi Dharmavaram is also with USA-CERL. The AFESC-ESL Project Officer was Surendra B. Joshi.

This report was prepared by the U.S. Army Construction Engineering Research Laboratory Environmental Division (USA-CERL-EN) and U.S. Air Force Engineering and Services Center, Engineering and Services Laboratory (AFESC-ESL). Dr. R.K. Jain is Chief, EN, and LTC Thomas J. Walker is Chief of the AFESC-ESL Environics Division. The technical editor was Dana Finney, USA-CERL Information Management Office.

COL Carl O. Magnell is Commander and Director of USA-CERL, and Dr. L.R. Shaffer is Technical Director. COL Roy G. Kennington is the Commander of AFESC.

This work was performed for the U.S. Army Engineering and Housing Support Center (USAEHSC); the USAEHSC technical monitor was Thomas Wash.

NOTICE

Contents and Subject Index

PART II
VAPOR DEGREASING AND PRECISION CLEANING SOLVENTS

Part I

Cold-Cleaning Solvents

The information in Part I is from *Used Solvent Testing and Reclamation, Volume I: Cold-Cleaning Solvents,* prepared by Arthur R. Tarrer, Auburn University; Bernard A. Donahue and Seshasayi Dharmavaram, U.S. Army Construction Engineering Research Laboratory; and Surendra B. Joshi, U.S. Air Force Engineering and Services Center; for the U.S. Army Engineering and Housing Support Center, December 1988.

1. Introduction

Background

Department of Defense (DOD) installations use large amounts of solvent each year in cleaning operations, which generates a huge volume of waste solvent. Much of this waste is or will be considered hazardous as stricter regulations are promulgated and enforced. Thus, proper handling and disposal practices are becoming of increasing concern to DOD. Coupled with these environmental issues is the rising cost of both waste disposal and new solvents. These concerns have prompted DOD to seek safe, cost-effective methods of managing waste solvent.

Solvents used at DOD installations can be classified into five groups based on chemical makeup and function: (1) vapor degreasers, (2) cold-dipping cleaners, (3) paint thinners, (4) paint strippers and carbon removers, and (5) precision cleaners. Most of these solvents are considered to provide one-time use; when they become contaminated, they are discarded. In these cases, disposal methods are mainly destructive, i.e., waste solvents are incinerated, evaporated, or dumped.

Some military facilities have initiated programs for reclaiming used solvents. This option is technically feasible because the solvents usually do not break down chemically during cleaning operations. Their role in cleaning is limited mainly to physical solubilization of waxes, greases, oils, and other contaminants. In fact, laboratory tests of major waste streams at installations have indicated that most solvents present could be recovered by recycling; the reclaimed material would generally be of suitable quality for effective reuse in cleaning.

A situation that has limited recycling is the lack of scientific tests and criteria for judging a solvent as spent (i.e., contaminated to the point that it is no longer effective for its intended purpose). Discarding a solvent before this point fails to maximize the material's life from an economic standpoint, whereas keeping it in service too long may result in use of an ineffective cleaner.

To make solvent recycling practical at installations, DOD needs criteria and simple test(s) for identifying spent solvents and/or indicating when the solvent should be discarded. These tests could have major impact on the environmental and cost issues facing DOD. Specifically, effective tests could:

1. Maximize solvent life for the most economical use of product; solvent cleaning operations would realize a savings through a reduction in new purchases.

2. Allow recycled solvent to be evaluated, preventing the use of inadequate quality materials in the cleaning process.

3. Minimize the amount of hazardous wastes generated, thus limiting the cost of handling and disposal. This benefit is especially important in light of the DOD Used Solvent Elimination (USE) program,[1] which bans the disposal of used solvent in landfills.

Purpose

The overall purpose of this work is to (1) establish criteria for identifying spent solvents and recommend simple tests to determine when solvents must be changed and (2) evaluate methods of reclaiming solvents as an alternative to disposal. Volume I addresses solvents used mainly in cold-cleaning operations (e.g., mineral spirits, PD-680). Volume II covers halogenated solvents used in vapor degreasing and metal cleaning/surface preparation.

Approach

This phase of the study involved the following steps:

1. Review the literature for methods of monitoring solvent quality.

2. Investigate physicochemical test methods.

3. Analyze Stoddard solvents (SS) and evaluate test results.

4. Rate the test methods and determine their practicality for use in the field.

5. Evaluate methods of reclaiming spent and partially spent solvent.

6. Study the economics of solvent reclamation.

7. Recommend test methods and reclamation strategies for use at military installations.

Mode of Technology Transfer

It is recommended that the test procedures be verified in the field and refined through a transfer medium such as the Facilities Engineering Applications Program (FEAP) program. When the tests are validated, they should be implemented at all military installations where solvents are used. The procedures would be incorporated into the appropriate technical manuals for implementation.

---

[1]Memorandum from the Office of the Assistant Secretary of Defense, Director of Environmental Policy, "Used Solvent Elimination (USE) Program," Interim Guidance (February 1985).

2. Solvent Properties and Use Within DOD

Cleaning Mechanism

For background information, the literature was surveyed to learn about solvent chemistry and its role in the overall cleaning operation. To understand how a solvent works, it is first necessary to study the types of contaminants that require solvent cleaning. Solvent contaminants are generally a heterogeneous mixture of substances with different physical and chemical characteristics. They can be roughly grouped as:

1. Hydrocarbon oils, such as lubricating oils and greases, transmission oil, fuel oil, asphalt, and tar.

2. Paints and varnishes.

3. Soily material, such as clay and silt, cement, soot, and lampblack.

The first two categories are inert organic materials (liquids and semiliquids), whereas the third category consists mainly of insoluble inorganic materials which are, as a group, solids in various states of subdivision. However, the contaminants can occur as liquid, solid, or a combination of the two (e.g., soot in oily material).

Several phenomena contribute to adhesion of these contaminants to metal parts, all of which can be classified generically as: (1) mechanical entrapment onto/into the part; (2) bonding to the part by cohesion or wetting; and (3) bonding by chemical or adsorptional combination with the part material.

The first category, mechanical entrapment, is limited to solid soil or contaminants on the rough protuberances of a part or in crevices of the part structure.

The second mechanism of attachment is by cohesion or wetting. Here, the contaminant, if in liquid form, is held by wetting of the part surface; in the case of a liquid/solid mixture, the liquid acts as an adhesive between the solid particles and the part material by mutual wetting. The bond can be broken by an increase in the free energy of the liquid contaminant/part interface, substitution of a new interface that has lower free energy (by adding surfactants), or imparting of mechanical energy to the contaminant/part interface as by brushing or wiping action.

The third mechanism, chemical or adsorptional combination of the contaminants with the part material, is not generally observed in routine cleaning operations except in situations for which a part has been exposed to strong acids or alkalis. Thus, this report emphasizes the first two mechanisms.

Given these mechanisms, an effective cleaning operation can be divided into three steps:

1. Separation of contaminants from part material.

2. Dispersion and solubilization of contaminants in the solvent.

3. Stabilization of the dispersed contaminants.

4

In many situations, these steps are interdependent and the limits of each are not always defined clearly.

The known actions of solvents that characterize their ability to effect cleaning are: (1) wetting and spreading on both part and contaminant surfaces, (2) lowering the surface free energy of contaminant surfaces, (3) speeding movement of solid and liquid particles, and (4) solubilization. The degree of solubility of a solute in a solvent is known as the "solvent power" of the solvent. For a homologous series of hydrocarbons, the volume of a liquid solute (oil and grease) solubilized at constant temperature is generally inversely proportional to the molar volume of the solute. Thus, the degree of solubilization depends to a large extent on the chain length of a hydrocarbon solute. Greases of high molecular weight are sparingly soluble in SS. The situation is complicated by the fact that solvents such as SS and mineral spirits are themselves a mixture of aliphatic and aromatic hydrocarbons. Different groups or series of hydrocarbons in SS have different solubilities for oily and greasy contaminants.[2]

Although solubilization has been studied by many researchers, most of the literature relates to final equilibrium solubilization of oils by surfactants. Only a few studies are concerned with the mechanism and kinetics.[3]

Stabilization of insoluble materials means that solid and liquid contaminants are kept dispersed in the cleaning medium to minimize redeposition. (Most studies in this area have been limited to aqueous surfactants acting on soiled fabrics.) One method of decreasing redeposition is by subjecting a part to multiple dippings in different vats instead of one-time dipping in a single vat. In the first vat, contaminants are removed from the part up to a certain point in a short time. This vat thus provides preliminary cleaning; a complete cleaning consists of two or more of these short dippings in additional vats. The solvent can be cycled from vat to vat on a periodic basis counter to the flow of parts, i.e., solvent from a "cleaner" vat could be moved to a preceding vat where a lower degree of cleaning is required. Solvent from the first vat (preliminary cleaning) would be reclaimed or discarded after it is spent. The number of vats in the counterflow scheme need not be large--two or three would be enough for routine cleaning.

Another method of decreasing redeposition is to use filters that trap insoluble particulates. This option leaves the vat liquid essentially free of solids and insoluble mass.

---

[2]W. W. Niven, *Fundamentals of Detergency* (Reinhold, 1950); J. M. Rosen (Ed.), Structure/Performance Relationships in Surfactants, ACS Symposium Series 253 (American Chemical Society [ACS], 1984).
[3]J. M. Rosen (Ed.), *Structure/Performance Relationships in Surfactants*, ACS Symposium Series 253 (ACS, 1984).

Solvent Characterization

Several characteristics are required of solvents to be used in the cold degreasing process.[4] They must:

- Dissolve oils, grease, and other contaminants.

- Have a high vapor density relative to air and a low vapor pressure to minimize solvent losses.

- Be chemically stable under conditions of use.

- Be essentially noncorrosive to common construction materials.

- Have a boiling point and latent heat of vaporization low enough to permit the solvent to be separated easily from oil, grease, and other contaminants by simple distillation.

- Not form azeotropes with liquid contaminants or with other solvents.

- Remain nonexplosive and nonflammable under the operating conditions.

A typical solvent satisfying these properties is mineral spirits. This type of solvent is known by various trade names, e.g., SS, drycleaning solvent, PD-680, Varsol (Exxon), turpentine substitute, white spirit, and petroleum spirit. SS is used mainly in cold cleaning metal parts and equipment in vats. Its low evaporation rate and high flashpoint make it a widely used solvent for removing oils, greases, and dirt from metal. Table 1 lists the specifications of Types I and II SS as developed by the American Society for Testing and Materials (ASTM).[5] The two types are distinguished primarily by their flashpoints, i.e., Type I solvent flashes at 100 °C whereas Type II flashes at 140 °C. Since these solvents are petroleum distillates, variations in solvent power between batches can be significant as reflected by Kauri-butanol values (between 29 and 45).

There is a broad distillation range for these solvents and the components are mainly alkane isomers between C_6 and C_{16}. They contain approximately 2 percent toluene and a maximum of 0.5 percent benzene. The toxic hazard rating for mineral spirits is considered to be slight to moderate.[6]

---

[4] *Source Assessment: Reclaiming of Waste Solvents, State of the Art*, EPA 60012-78-004f (U.S. Environmental Protection Agency [USEPA], April 1978); *Source Assessment: Solvent Evaporation-Degreasing Operations*, EPA 600/2-79-019f (USEPA, August 1979).

[5] *Annual Book of ASTM Standards*, Vol 6.03, Section 6 (American Society for Testing and Materials [ASTM], 1983).

[6] A. L. Bunge, *Minimization of Waste Solvent: Factors Controlling the Time Between Solvent Changes*, USA-CERL Contract No. DACA 88-83-C-0012 (Colorado School of Mines, September 1984).

Table 1

Physical and Chemical Properties of Stoddard Solvent*

Property	Type I	Type II
Commercial reference	Stoddard	140°F Solvent
Appearance	Clear and free of suspended matter and undissolved water	
Flashpoint, min. °F (°C)	100 (38)	140 (60)
Distillation, °F (°C)		
Initial boiling pt., min.	300 (149)	350 (177)
50% recovered, max.	350 (177)	385 (196)
Dry point, max.	407 (208)	412 (211)
Kauri-butanol value:		
Min.	29	29
Max.	45	45
Color, max.	Water-white or not darker than +2	
Residue from distillation:		
Percent, max.	1.5	1.5
Acidity	Neutral	Neutral
Doctor test	Negative	Negative
Copper corrosion		
Max. rating	1	1
Odor	Characteristic, nonresidual	
Bromine no., max.	5.0	5.0
Apparent specific gravity (60°F/60F):		
Min.	0.754	0.768
Max.	0.820	0.820

*Source: *Annual Book of ASTM Standards*, Vol 6.03 (ASTM, 1983). Used with permission.

Generation of Waste Stoddard Solvent

Some 485,000 cold cleaning operations in the United States use Stoddard-type solvents. The annual consumption of SS in degreasing operations amounts to more than 220,000 MT.[7]

In assessing SS usage within DOD, the military installations can be classified as large and small.[8] Large bases include shipyards, air logistics centers, Army depots, and air rework facilities. Each installation in this category annually generates 500 drums (55 gal) of SS, and there are 29 such installations. Small installations are much more

[7]Source Assessment: *Reclaiming of Waste Solvents, State of the Art; Source Assessment: Solvent Evaporation-Degreasing Operations.*
[8]R. W. Bee and K. E. Kawaoka, *Evaluation of Disposal Concepts for Used Solvents at DOD Bases*, TOR-0083(3786)-01 (February 1983).

numerous (at least 124 bases) and are low-volume solvent consumers.[9] Representative annual volumes for small bases are 150 drums of SS each. Table 2 lists the amount of spent solvent observed at some large and small DOD installations.

A pollution abatement study conducted by Lee et al. revealed that a large volume of solvents also is being used annually by five of the six Naval Air Rework Facilities (NARFs).[10] Table 3 lists quantities used by each NARF.

Disposal Alternatives

The waste solvent generated at DOD installations is disposed of in four primary ways: incineration, fuel substitution, surface disposal, and sale. These practices were reviewed and are summarized below.

Incineration

Incineration is the burning of wastes with air in an incinerator. This disposal alternative is useful for a number of waste solvents including cold-dipping wastes. However, segregation of these wastes from chlorinated hydrocarbons, lead and other metals, and water contaminants is essential so that emissions from the incinerator satisfy U.S. Environmental Protection Agency (USEPA) and state regulations. Incineration is an attractive disposal method for wastes that contain a variety of hydrocarbon products that cannot be reclaimed readily. Incinerators have been used to dispose of a variety of oil and grease waste generated at refineries.[11] In some instances, effluent combustion gases from this burning are passed through heat exchangers to preheat process streams.

Fuel Substitution

Many DOD installations burn spent or waste solvents in boilers and other firing equipment. This option is preferable to incineration because the fuel energy of the waste solvent is used beneficially. However, the wastes have to be segregated such that no halogenated hydrocarbons can contaminate the waste streams.

Several industrial studies have shown that blending waste solvents with regular fuel causes no noticeable increase in burner maintenance costs (due to corrosion) or adverse air pollution effects.[12] The advantages to combustion are that liquid wastes are disposed of profitably and safely, resulting in conservation of solvent resources.

Surface Disposal

Heavy nonvolatile waste solvents have been used in the past as road oil for dust control. However, waste streams consisting primarily of SS would evaporate into the air or run off with rain, thus contaminating local water sources (e.g., streams and lakes).

---

[9]R. W. Bee and K. E. Kawaoka.

[10]H. J. Lee, I. H. Curtis, and W. C. Hallow, *A Pollution Abatement Concept, Reclamation of Naval Air Rework Facilities Waste Solvent, Phase I* (Naval Air Development Center, April 1978).

[11]*Disposal/Recycle Management System Development for Air Force Waste Petroleum Oils and Lubricants,* AD779723 (U.S. Air Force, April 1974).

[12]*Disposal/Recycle Management System Development for Air Force Waste Petroleum Oils and Lubricants.*

Table 2

Spent Solvent Generation at Major DOD Bases*

Bases	Cleaning Bath Solvents (55-gal drums/yr)	All Solvents Total (55-gal drums/yr)
Seneca Army Depot, NY	140	180
Kelly AFB, TX	152	1134
Hill AFB, UT	455	2270
Tyndall AFB, FL	50	118
Jacksonville NAS, FL	60	2285
Davis Monthan AFB, AZ	190	227
Bergstrom AFB, TX	170	243
Corpus Christi Army Depot, TX	750	1025
Norfolk NARF, VA	540	1084
McClellan AFB, CA	90	935
Anniston Army Depot, AL**	660	1455

*Source: R. W. Bee and K. E. Kawaoka, *Evaluation of Disposal Concepts for Used Solvent at DOD Bases*, TOR 0083(3786)-01 (The Aerospace Corp., February 1983). Used with permission.
**Survey by the Chemical Engineering Dept., Auburn University.

Table 3

Spent Solvent Generated at Naval Air Rework Facilities*

Base	Stoddard Solvent Generation Rate (1000 gal/yr)	All Solvents (1000 gal/yr)
Alameda, CA	90.0	176
Norfolk, VA	24.0	78.1
North Island, CA	262.9	392.8
Pensacola, FL	57.2	209.9
Jacksonville, MI	7.7	106.4
Total	450.9	963.2

*Source: H. J. Lee, I. H. Custis, and W. C. Hallow, *A Pollution Abatement Concept, Reclamation of Naval Air Rework Facilities Waste Solvent, Phase I* (Naval Air Development Center, April 1978).

Another method of surface disposal is by contacting waste solvents with oil-consuming soil microorganisms under ambient conditions.[13]   Results using this method have been encouraging, except in cases of excessive rainfall which led to runoff of hydrocarbons into adjacent water sources.

Sanitary landfilling has been selectively used for disposal of petroleum-based wastes, usually in combination with other landfilled waste. Use of this mechanism for disposal of solvent wastes is still experimental.

Sale

The Defense Reutilization and Marketing Service (DRMS) is responsible for selling used solvent material. The sale option is profitable as well as an environmentally acceptable disposition practice, provided the buyers use appropriate reprocessing methods. The buyers' qualifications must be evaluated carefully because unscrupulous buyers have been known to improperly release the used materials into the environment, which could result in potential DOD liability. Used solvent material for sale must be segregated at the point of generation and labeled properly. Mineral spirit wastes have been sold at some installations for $0.40/gal. [14]

[13]H. Kobayashi and B. E. Rittman, "Microbial Removal of Hazardous Organic Compounds," *Environ. Sci. Technol.*, Vol 16, No. 3 (1982).

[14]R. W. Bee and K. E. Kawaoka; *Disposal/Recycle Management System Development for Air Force Waste Petroleum Oils and Lubricants.*

3. Evaluation of Detection Methods

Literature Review

A literature survey indicates extensive use of SS as a cleaning agent in the dry-cleaning industry.[15] Although there is a standard for new or virgin SS (ASTM D 484-71), no such specification is available for spent or contaminated solvents. However, some arbitrary criteria are used in various industries; for example, in the drycleaning industry, the transmittance of light through a sample of dirty solvent is the determining factor as to when to change the solvent.[16] In other degreasing operations, the solvent color as well as the presence of dirt are taken into consideration for changing these agents.[17]

Various physical properties of solvents have been proposed to reflect the extent of contamination. These include boiling point, density, refractive index, color, and viscosity.[18] However, these properties may not be truly indicative of a solvent's ability to remove grease and soil. In addition, variations in the values of these properties, as a function of solvent strength, may be small and thus require very accurate measurements. This factor could be a potential problem for a test that is to be performed at a field station, although simple modern instrumentation may facilitate such measurements.

Color is commonly used as a criterion to evaluate dirty solvents in the drycleaning industry.[19] Generally, a clean solvent will allow a transmission of 90 percent visible radiation at 500 nm. However, after prolonged use, the transmittance drops to the vicinity of 50 percent (after filtration of solid residue), and the solvent is discarded for reclamation. The addition of aqueous hydrogen peroxide to dirty solvents rejuvenates them and allows reuse of vat solvents without distillation. It is speculated that a dual action occurs: (1) bleaching by the peroxide which decreases light absorption and (2) a physical effect on the suspended fine dirt by particles of aqueous hydrogen peroxide; these particles act as nuclei to which small dirt particles agglomerate to a size where they will filter out of the solvent.[20]

---

[15]*Source Assessment: Reclaiming of Waste Solvents, State of the Art; Source Assessment: Solvent Evaporation-Degreasing Operations; Annual Book of ASTM Standards*; A. E. John, *Drycleaning* (Merrow Publishing Co., England, 1971); A. R. Martin and G. P. Fulton, *Drycleaning Technology and Theory* (Textile Book Publishers, 1958); E. R. Phillips, *Drycleaning* (National Institute of Drycleaning, Inc., 1961); C. B. Randall, *The Drycleaning Department* (National Association of Dyers and Cleaners, 1937); K. Johnson, *Drycleaning and Degreasing Chemicals and Processes* (Noyes Data Corp., 1973); I. Mellan, *Industrial Solvents Handbook* (Noyes Data Corp., 1977); T. H. Durrans, *Solvents*, 2nd ed. (Van Nostrand, 1931); I. Mellan, *Industrial Solvents* (Reinhold, 1950); L. Scheflan and M. B. Jacobs, *The Handbook of Solvents* (Van Nostrand, 1953); *International Fabricare Institute Bulletin* T-447 (1969).

[16]K. Johnson; *International Fabricare Institute Bulletin; National Institute of Drycleaning Bulletin Service*, T-413 (1965); H. M. Castrantas, R. E. Keay, and D. G. MacKellar, *Treatment of Dry Cleaning Baths*, U. S. Patent 3,677,955 (July 1972).

[17]A. L. Bunge.

[18]A. L. Bunge; K. Johnson.

[19]K. Johnson.

[20]W. W. Niven; K. Johnson; *National Institute of Drycleaning Bulletin Service*.

The viscosity of a solvent usually increases greatly when the solvent becomes contaminated with grease and/or oil. This change is due to a high intermolecular attraction between the solvent and the dissolved contaminants.

In thin-layer chromatography, a standard dye is "spotted" onto a glass microfiber sheet, and the sheet is then placed in the solvent sample. As the solvent moves up the plate by capillary action, the dye migrates at a rate characteristic of the sample's solvent power. The distance traveled by the dye is divided by the distance traveled by the solvent to obtain the R_f value of the dye. The R_f value has been shown to have direct correlation with solvency and solvent power.[21]

Another category of tests (compared with those based on physical parameters) rates solvent power in terms of chemical reactivity or degree of solubility of certain materials. These properties include acid number, aniline point, dimethyl sulfate value, Kauri-butanol value (KBV), and cellulose-nitrate solution value.[22] KBV has been used traditionally in the drycleaning and varnish industries to represent solvent performance. Solvent KBV decreases with increased solvent contamination by grease, oil, or soil. A drawback to this test is that KBV reflects the solubility of Kauri resin in a solvent and possibly will not accurately represent the solvent's ability to dissolve compounds that have structures significantly different from those of Kauri resin.

Table 4 summarizes some current techniques for determining solvent quality.

Current Solvent Testing at Military Installations

To identify existing criteria for determining the change schedule of solvents at various DOD installations, a questionnaire was prepared and sent to 20 military facilities. However, responses were received from only three facilities. Of these three, Robins AFB, GA, does not have a cold-dipping operation that uses SS. SS is used only to clean grime and dirt from machinery parts. The Appendix presents the questionnaire in full along with responses.

Anniston Army Depot requires that all cleaned parts be tested by wiping with a clean white cloth. If there is no visible residue on the cloth, the part is considered to be clean. In addition, a laboratory centrifugation test is performed on the spent solvent. The solvent is changed when the amount of solids exceeds 2 percent.

Tyndall AFB, FL, sent responses from two operating departments. The EMS wheel and tire shop uses SS for removing carbon deposits, grease, and dirt from aircraft wheels and wheel bearings. The degree of cleanliness is determined through close inspection by experienced personnel. The 25th EMS AGE branch uses cold dipping to remove grease, dirt, etc., from parts prior to disassembly for repair or overhaul. The parts, as in the other department, must be very clean and the degree of cleanliness is determined by visual inspection.

---

[21]G. G. Esposito, Solvency Rating of Petroleum Solvents by Reverse Thin-Layer Chromatography, AD753336 (Aberdeen Proving Ground, 1972).

[22]*Annual Book of ASTM Standards*; A. L. Bunge; *International Fabricare Institute Bulletin.*

Table 4

Measurement Techniques for Determining Solvent Quality as Reported in the Literature

Method	Principle	Reference*
Electrical Conductivity (ASTM D 2624 & D 3114)	Measures the ability of a solution to carry an electric current. A metal part is cleaned with solvent and then dipped in a relatively more conductive liquid (e.g., ethyl alcohol). If the part is clean, the alcohol will not show a significant change in conductivity. Apparatus: conductivity meter and probe.	1 - 3
Ultraviolet Spectroscopy (ASTM D 1319)	Absorptivity of liquids at specified wavelength in the UV region is used to characterize petroleum products. Apparatus: UV spectrophotometer.	1 - 3
Fluorescent Indicator Adsorption (ASTM D 1319)	Sample is introduced into a glass adsorption column packed with activated silica gel. When the sample has been adsorbed on the gel, alcohol is added to desorb and force it down the column. The hydrocarbons are separated into aromatics, olefins, and saturates. A dye is added to the silica gel which illuminates the different zones under UV light. Apparatus: glass adsorption column.	4
Sediment Extraction (ASTM D 473)	Sample, contained in a refractory thimble, is extracted with hot toluene until the residue reaches constant mass. The mass of residue is reported as "sediment by extraction."	4
Thin-Layer Chromatography	Standard dye is spotted on a glass microfiber sheet and the edge of the sheet is placed into the solvent. After the solvent has migrated to the top of the sheet, the sheet is removed and dried. The ratio of the distances traveled by the dye and solvent is established and related to solvent power.	5,6
Density, Specific Gravity, or API Gravity (ASTM D 1298)	Sample is brought to a prescribed temperature. A hydrometer is placed into the sample container and allowed to come to equilibrium. The scale is read and converted to appropriate units. If a density meter or a specific gravity meter is used, the sample is brought within $5°C$ of ambient temperature after introduction into the meter oscillator and a digital readout is obtained. Apparatus: hydrometer or density meter.	4,7
Refractive Index	Measured by critical angle method with a refractometer. Refractive index is used to characterize hydrocarbons and their mixtures. Apparatus: refractometer, thermometer.	4,8

*References are listed in numerical order at the end of this table.

Table 4 (Cont'd)

Method	Principle	Reference
Acid Number	Determines the amount of fatty acid in a solvent. Defined as "mg of KOH to neutralize 1.28 mL of the solvent." Apparatus: buret, pipet, flask. Chemicals: KOH, methanol, phenolphthalein.	5,9,10
Color (Visible Absorbence)	Sample's light transmittance is observed between 450 and 600 nm. In drycleaning industry, if transmittance drops to near 50% after filtration, solvent is usually changed. Equipment: Visible spectrometer or colorimeter.	11,12
Kauri-Butanol Value (ASTM D 1133)	Gives an index number for ranking solvents. The basis of the test is that Kauri gum is very soluble in butanol but its solubility decreases as the butanol is diluted with a solvent that will not dissolve the resin. Apparatus: buret, flask, precision balance. Chemicals: Kauri-butanol solution.	6,7,9
Aniline Point (ASTM D 611)	Useful in characterizing pure hydrocarbons (HCs) and mixtures. Aromatic HCs exhibit the lowest, and paraffins the highest, values. Aniline point is most often used to estimate aromatic HC content of mixtures. Apparatus: aniline point apparatus, temperature bath, and thermometers. Reagents: aniline, $CaSO_4$, n-heptane, and Na_2SO_4.	9
Cellulose Nitrate Dilution Ratio for Active Solvents (ASTM D 1720)	Determines the volume ratio of HC diluent to active solvent required to cause precipitation in a solution of celluose nitrate. By using a standard material for any two of the three components, the effect of different types of third component can be determined. Apparatus: titration buret and accessories. Reagents: n-butyl acetate, cellulose nitrate, and toluene.	9

[1] *Annual Book of ASTM Standards*, Section 5, Vol 5.02 (1983).

[2] *Annual Book of ASTM Standards*, Section 5, Vol 5.03 (1983).

[3] H. H. Bauer, G. D. Christian, and J. E. O'Reilly, *Instrumental Analysis* (Allyn and Bacon, 1979).

[4] *Annual Book of ASTM Standards*, Section 5, Vol 5.01 (1983).

[5] *National Institute of Drycleaning Bulletin Service*, T-413 (1965).

[6] G. G. Esposito, *Solvency Rating of Petroleum Solvents by Reverse Thin-Layer Chromatography*, AD-753336 (Aberdeen Proving Ground, 1972).

[7] A. L. Bunge, *Minimization of Waste Solvent: Factors Controlling the Time Between Solvent Changes*, CERL Contract DACA 88-83-C-0012 (Colorado School of Mines, September 1984).

[8] *Refractometer Manual*, ABBE-56 (Bausch and Lomb Optical Co.).

[9] *Annual Book of ASTM Standards*, Section 6, Vol 6.03 (1983).

[10] A. R. Martin and G. P. Fulton, *Drycleaning Technology and Theory* (Textile Book Publishers, 1958).

[11] K. Johnson, *Drycleaning and Degreasing Chemicals and Processes* (Noyes Data Corp., 1977).

[12] H. M. Castrantas, R. E. Keay, and D. G. MacKellar, *Treatment of Dry Cleaning Baths*, U. S. Patent 3,677,955 (July 1972).

Anniston Army Depot was visited to observe the cold-dipping facilities there. Cold-dipping operations are done to clean transmission parts of tanks and military vehicles. A typical vat contains 50 to 60 gal of solvent. The parts are immersed in the solvent for a period of time and, occasionally, a brush is used to dislodge grime and dirt.

Anniston personnel were questioned as to how the samples are collected for determining the amount of solid and water contamination. According to these individuals, a layer of sludge usually forms on the bottom of a vat if a filtering system is not used. The sludge can easily be removed using a paper filtering system.

Fort Benning, GA, uses a portable 1-micron paper filtering system at its weapons consolidation pool. This filtering system has saved about $70,000 over just 6 months. In this operation, the solvent is mixed with an oil and other additives, and the mixture is used to wash weapons. This is a different type of cleaning operation than the others surveyed, but the savings are impressive and suggest that such filtering systems should be examined for use in similar applications.

Physicochemical Tests

To evaluate some of the physicochemical test methods, an experimental study was done using SS. The following properties were measured: (1) Kauri-butanol value, (2) viscosity, (3) specific gravity, (4) refractive index, (5) visible absorption, (6) electrical conductivity, and (7) thin-layer chromatography. These techniques were selected from the literature on the basis of reported scientific reliability and consistency, as well as relative ease of performing in field conditions. The background and procedures for these tests are described below.

Kauri-Butanol Value

As mentioned, the ASTM Kauri-butanol method is a convenient way to measure the relative power of solvents. This method includes an index for ranking solvents for their ability to dissolve other materials.[23]

SS is a mixture of hydrocarbons distilled from petroleum. It can vary in composition, depending on the nature of the petroleum from which it was distilled. SS contains three classes of hydrocarbons: aliphatic (paraffins), alicyclic (naphthenes), and aromatics (benzene and its derivatives). The paraffinic hydrocarbons have the lowest solvent power and the aromatics have the highest. Therefore, the solvent power in terms of Kauri-butanol value (KBV) largely depends on the amount of aromatics present. This is because the Kauri resin is standardized against (1) toluene (an aromatic compound) which has an assigned value of 105 and (2) a mixture of 75 percent heptane and 25 percent toluene (an aliphatic-aromatic mix) which has an assigned value of 40. The basis for the test is that the Kauri resin becomes less soluble in "weaker" solvents such as aliphatic hydrocarbons. Therefore, the solvent being tested is added in small portions until the solution becomes cloudy due to precipitation of the resin. The greater the volume of solvent added to the Kauri-butanol (K-B) solution before it becomes cloudy, the higher the KBV. This simple test is widely accepted as a good measure of relative solvent power. The methodology is as follows.[24]

[23] *Annual Book of ASTM Standards; National Institute of Drycleaning Bulletin Service;* G. G. Esposito.
[24] *Annual Book of ASTM Standards.*

Apparatus:

- Erlenmeyer flask, 250-mL.

- Buret, 50-mL (Figure 1).

- Print sample (a sheet of white paper with black 10-point print, No. 31 Bruce Old Style type).

Reagents:

- Standard K-B solution; a prepared solution was obtained from Chemical Service Lab, 5543 Dyer St., Dallas, TX 75206.

- Reagent-grade toluene.

- Reagent-grade heptane.

Procedure:

1. Weigh 20 g of standard K-B solution into a 250-mL Erlenmeyer flask.

2. Place the sheet of 10-point print under the flask.

3. Fill buret with solvent to be sampled.

4. Titrate into the flask until printed material becomes obscured or blurred but not to the point where the print becomes illegible.

5. Calculate the Kauri-butanol value using the following formula:

$$KBV = 65(C-B)/(A-B) + 40 \qquad \text{[Eq 1]}$$

where:

A = mL of toluene required to titrate 20 g of K-B solution (should be around 105).

B = mL of 75 percent heptane/25 percent toluene blend needed to titrate 20 g of K-B solution (should be around 40).

C = mL of sample solvent needed to titrate 20 g of K-B solution.

Values of A and B can be obtained from standard titrations or from the supplier of the K-B solution.

Viscosity

Viscosity is the internal friction or resistance to flow that exists within a fluid, either liquid or gas. This property depends on the intermolecular attractive forces within the fluid. Viscosity is an extremely useful method for characterizing oils and solvents. Viscosities of "heavier" and "lighter" oils are significantly different, whereas their densities may differ very little.

Figure 1. Kauri-butanol value apparatus.

A common unit of viscosity is the poise, which is equal to 1 gram per centimeter second (g/cm-sec), usually given in centipoise (cp).   Viscosities in this work were measured using an Ostwald viscometer. This type of viscometer measures the flow rate of a fluid through a capillary tube in a gravity field. Newtonian behavior was assumed for all solvent mixtures.   A Newtonian fluid is one that shows a linear relationship between the magnitude of the applied shear stress and the resulting rate of deformation. The viscosity (μ) of a given fluid is calculated using Equation 2:

$$\mu = k \times t \qquad\qquad\qquad \text{[Eq 2]}$$

where t is the time required for a fixed volume of fluid to flow through the capillary and k is a constant obtained by measuring the time of a liquid that has a similar known viscosity.

Apparatus:

● Ostwald viscometer (Figure 2).

● Stopwatch.

Figure 2. An Ostwald viscometer.

Procedure:[25]

1. Wash the viscometer thoroughly and rinse with distilled water, making sure the instrument is clean and dry before taking readings.

2. Introduce distilled water and allow it to come to thermal equilibrium in a constant-temperature bath.

3. Draw liquid into upper bulb to marked line with a suction bulb.

4. Remove bulb and record time needed for level of water to pass between markings.

5. Repeat steps 3 and 4 until readings are fairly constant.

6. Clean and dry viscometer thoroughly.

7. Add an appropriate volume of the solvent to be tested into the viscometer.

8. Repeat steps 3 through 5.

Refractive Index

The refractive index of a liquid is the ratio of the velocity of light in a vacuum to the velocity of light in the liquid. This property can be used to identify a substance and determine its purity. Since the angle of refraction varies with the wavelength of light

[25]G. J. Shugar, et al., *Chemical Technician's Ready Reference Handbook,* 2nd ed. (McGraw-Hill, 1981).

used, the measurement of refractive index requires that light of a known wavelength be used. However, a white light can be used if the refractive index of a reference liquid is measured in the same light.

Refractive index is commonly reported to four decimal places and, since it can easily be determined experimentally to a few parts in 10,000, it is a very accurate physical constant. Small amounts of impurities can have significant effects on the experimental value. Refractive indices in this study were determined using an Abbe refractometer.[26] This instrument compares the angles at which light from an effective point source passes through a test liquid and into a prism for which the refractive index is known.

Procedure:[27]

1. Unlock the hinged assembly and lower the bottom part of the prism.

2. Clean the upper and lower prisms with soft, nonabrasive, absorbent, lint-free cotton wetted with benzene. Rinse by wiping with petroleum ether and allow to dry.

3. Place a drop of solution of known refractive index (e.g., water) on the prism.

4. Record the temperature indicated by the thermometer next to the prism.

5. Set the scale to correspond with the known refractive index at the corresponding temperature.

6. Look through the eyepiece and turn the compensator knob until the colored, indistinct boundary seen between the light and dark fields becomes a sharp line.

7. Adjust the magnifier arm until the sharp line exactly intersects the midpoint of the crosshairs in the image (Figure 3).

8. Repeat steps 1 through 7 using the solvent to be tested.

9. Clean prisms and lock them together.

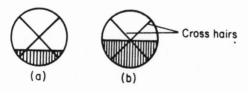

(a)   (b)

Figure 3. Adjusting the refractometer.

---

[26] *Refractometer Manual*, ABBE-56 (Bausch and Lomb Optical Corp.).
[27] *Refractometer Manual*.

Specific Gravity

Specific gravity is defined as the ratio of the density of a liquid to that of water at the same temperature. Density is a fundamental physical property of a substance denoting the mass of a substance per unit volume.

Specific gravity can be measured easily using a specific gravity meter, a hydrometer, or a pycnometer. Any of these devices can give very accurate results with little training or experience on the part of the tester. A hydrometer generally is not sensitive enough to detect the small specific gravity variations that occur when a solvent becomes contaminated.

In this study, specific gravity of SS was measured using two devices: (1) a pycnometer bottle, which holds a precise volume of liquid and is weighed on a balance, and (2) an electronic specific gravity meter (Mettler/Paar DMA35SG).[28] Both methods give accurate results, but the specific gravity meter is easier to use and requires no weighing. Specific gravity is very sensitive to changes in temperature because, as temperature increases, fluids have a tendency to expand, thus reducing the amount of mass in the same volume of fluid. The opposite effect occurs when a fluid is cooled.

Apparatus:

• Analytical precision balance.

• Pycnometer (Figure 4), or

• Electronic specific gravity meter.

Procedure:[29]

1. Clean pycnometer thoroughly.

2. Dry pycnometer in an oven for 30 min.

3. Remove from oven and allow to cool.

4. Weigh pycnometer on a precision balance.

5. Repeat steps 2 through 4 until a constant weight is obtained.

6. Fill pycnometer completely with liquid.

7. Wipe cap with tissue and weigh pycnometer.

---

[28]*Mettler/Parr DMA 35 Density Meter* (Mettler Instrument Corp., 1986).
[29]G. J. Shugar, et al.

Figure 4. Pycnometer.

When using a specific gravity meter:[30]

1. Turn on meter.

2. Fill bulb on meter, making sure no air bubbles are in the measuring tube as this will cause errors.

3. Record the temperature of the liquid as well as the specific gravity referenced to 20 °C.

4. Turn off meter.

5. Empty meter of all fluid and clean thoroughly.

Electrical Conductivity

According to Ohm's Law, the resistance of a conductor of length L and cross sectional area A is given by:

$$R = k \times (L/A) \qquad\qquad \text{[Eq 3]}$$

where R is the resistance in ohms and k is the specific resistivity, an inherent property of the material being examined which is expressed in ohm-cm. In dealing with liquids, the reciprocal of k usually is measured; this value is called the "specific conductance" or "conductivity" and is expressed in $ohm^{-1}cm^{-1}$ or mhos/cm. Thus, from Equation 3, the conductivity is given by:

$$c = (L/A) \times (1/R) \qquad\qquad \text{[Eq 4]}$$

---

[30]*Mettler/Parr DMA 35 Density Meter.*

Electrical conductivity is generally measured using a conductance cell for which the factor L/A in Equation 4 can be determined by measuring the known conductance of a standard solution, usually potassium chloride in water. For a cell of given geometry, the factor L/A is called the "cell constant" and, once it has been determined, the conductance of unknown solutions can be determined by applying the same procedures.

The electrical conductivity of pure organic liquids is usually very small--on the order of 10^{-8} mhos/cm or less at 25 °C. Because electrical conductivity is a function of temperature, some attention must be given to controlling the liquid's temperature while determining this value.

Since the conductance of a solution is a function of concentration, it would be expected that the electrical conductivity of a solvent, e.g., new SS, would change as impurities are accumulated during usage. Thus, monitoring a solvent's conductivity might provide one method of indicating solvent bath contamination. During the course of solvent usage, the conductivity could increase or decrease, depending on the particular impurities being accumulated.

Apparatus:

Conductivity meter (YSI Model 32).

Probe (YSI #3402).

Procedure:

1. Clean probe thoroughly (Figure 5).

2. Measure temperature of solvent to be tested.

3. Set conductivity meter to conductivity setting.

4. Dip probe in solvent and set instrument to proper scale.

5. Allow about 5 min for probe to reach equilibrium.

6. Record conductivity.

7. Repeat until constant.

Visible Absorbence

Because there is usually a noticeable change in the color of a contaminated solvent, light absorbence is an obvious choice for a test method. The amount of light absorbed could indicate the amount of contamination of a solvent. This type of test is currently used in the drycleaning industry.[31]

---

[31] K. Johnson; I. Mellan.

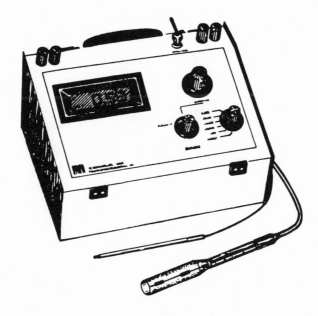

Figure 5. Conductivity meter with probe.

When an electromagnetic wave of a specific wavelength impinges upon a substance, the fraction of radiation absorbed will be a function of the concentration of the substance in the light path and the thickness of the sample. It has been found that increasing the concentration of the absorber has the same effect as a proportional increase in the radiation-absorbing path length. Therefore, the absorbence, A, is proportional to the concentration of absorbing solute:[32]

$$A = a \times b \times C \qquad\qquad\qquad \text{[Eq 5]}$$

where b is the sample path length (cm), a is specific absorptivity in $g^{-1}\text{-}cm^{-1}$, and C is the solvent concentration in g/L. This equation holds only for low concentrations. The derivation of Beer's law assumes the use of monochromatic light; however, if absorptivity is essentially constant over the instrumental bandwidth, Beer's law will be followed closely.   Departure from Beer's law is most serious for wide slit widths and narrow absorption bands, and is less significant for broad bands and narrow slits. Therefore, the most significant measurements were made using a very narrow slit width of 0.02 mm and a broad band between 400 and 600 nm.

[32]H. H. Bauer, G. D. Christian, and J. E. O'Reilly; H. H. Willard, L. L. Merritt, Jr., and J. A. Dean, *Instrumental Methods of Analysis*, 5th ed. (D. Van Nostrand, 1974).

Apparatus:

- UV/visible spectrophotometer (Gilford 250).

- Cuvette.

Procedure:

1. Turn on instrument and allow a 30-min warmup.

2. Clean and dry cuvette, making sure that all smudges are wiped off and there are no scratches on any surfaces in the light path.

3. Fill cuvette with sample and place in the spectrometer.

4. Set slit width and take reading.

Thin-Layer Chromatography

Chromatography, in general, is a separation technique based on the fact that a substance has different affinities for each of two phases--stationary and mobile. The relative distribution of a substance between the two phases is known as the distribution coefficient, K. The fact that different substances have different distribution coefficients makes it possible to separate them by chromatography. Two substances, A and B, with unequal distribution coefficients KA and KB will spend different periods of time in mobile and stationary phases. Movement of the mobile phase leads to a separation. For example, if KA > KB, then A spends more time in the mobile phase and thus travels faster than substance B.[33]

In classic thin-layer chromatography (TLC), a mixture to be separated (in this case, a mixture of dyes) would be deposited at a starting point on a plate or paper and the mobile phase (solvent) would be allowed to travel up the plate/paper by capillary action. Because of differences in distribution coefficients, the substances (dyes) in the mixture separate. The distances traveled by substances A and B, and the solvent (mobile phase), respectively, would then be recorded. Separation efficiency is presented as R_f (response factor) values, where R_f is the ratio of the distance traveled by a substance being separated to that of the solvent (mobile phase).

The ability of a solvent to keep a substance in solution can be termed its "solvent power" with respect to that substance. By visualizing a solvent as the mobile phase in TLC, it is clear that a solvent with higher solvent power should produce higher KA and R_f values than a solvent with lower solvent power. TLC has been used to rank the solvent power of various petroleum solvents.[34]

Apparatus:

- TLC glass microfiber paper.

- Dyes (Brilliant Blue and Disperse Yellow 9).

---

[33]G. G. Esposito; H. H. Bauer, G. D. Christian, and J. E. O'Reilly; H. H. Willard, L. L. Merritt, Jr., and J. A. Dean.
[34]G. G. Esposito.

Procedure:

1. Activate paper by heating it in an oven for 30 min.

2. Cut paper into strips that will fit into TLC jar.

3. Place a mark across paper approximately 2 cm from the bottom.

4. Pour solvent to be tested into jar to a level of about 1 cm.

5. Spot dye(s) on paper at the 2-cm mark.

6. Place paper into jar so that it remains vertical.   The dye spots should be approximately 1 cm above the liquid.

7. Cover the jar so that the liquid and vapor can come to equilibrium.

8. When the solvent (mobile) phase nears the top of the strip, remove strip from bottle and mark solvent front (Figure 6).

9. Allow strip to dry.

10. Measure the distance traveled by the dyes from the starting line to the middle of the dye spot. Measure the distance traveled by the solvent front. The ratio of the distance traveled by the dye to that of the solvent is the R_f value.

Stoddard Solvent Analysis and Test Results

Samples of new and spent SS were obtained in 5-gal containers from Anniston Army Depot, AL. From these samples, new samples were prepared (test series 1) using different proportions of new and spent solvents (0, 25, 50, 75, and 100 percent by volume of spent solvent). The properties measured for this set of samples were Kauri-butanol value, viscosity, specific gravity, refractive index, and visible absorption.

An arrangement was made with the quality control officer at Anniston Army Depot to obtain solvent samples from a vat at various time intervals prior to failure and removal of the solvent. The physicochemical properties measured for this series of time-study samples (test series two) provide the transient properties profile of the solvent. Nine periodic samples were received, including some taken the day the solvent was charged and the day it was discarded. The solvent usage life was 22 days. It was learned that when SS is charged to a vat (50 to 60 gal), 2 gal of motor lubricating oil are added to minimize drying of hands and other dermatological effects on the workers.

The new and spent SS samples were analyzed by gas chromatography (GC), and some of the many components present in these solvents were identified (see **Cleaning Mechanisms** above). Again, the physicochemical properties measured were:

- Kauri-butanol value.

- Viscosity.

- Refractive index.

- Specific gravity.

- Electrical conductivity.

- Visible absorbence.

- Thin-layer chromatographic profile.

The results of these measurements are described for two sets of conditions: (1) test series 1, i.e., laboratory simulation of solvent contamination by mixing various proportions of new and spent solvent (0, 25, 50, 75, and 100 percent by volume of spent solvent), and (2) test series 2, i.e., authentic time-study series samples.

Gas Chromatographic Analysis of Stoddard Solvent

An attempt was made to identify the primary constituents of SS using GC. The equipment used for this analysis was a Varian model 3700 gas chromatograph. All n-alkane constituents were identified. Figure 7 shows a typical chromatogram of new SS. This chromatogram contains many peaks; the four major n-alkane components were identified as decane, undecane, dodecane, and tridecane. The n-alkanes generally are higher boiling compounds than their respective isomers. This is true because n-alkanes have the greater hydrogen bonding due to lower steric hindrance of the molecules. Since most isomers of an alkane have lower boiling points than the n-alkane isomer, the series of peaks eluted prior to each n-alkane peak is speculated to be those of the branched isomers.

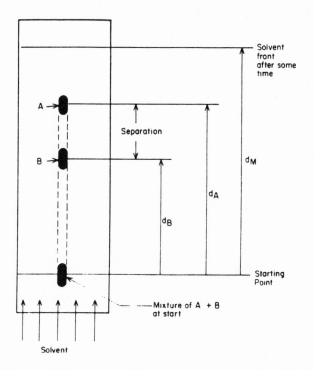

Figure 6. Mechanism of thin-layer chromatography.

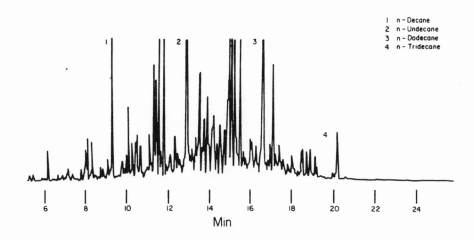

Figure 7.   New Stoddard solvent chromatogram. DB-5 Fused Silica Capillary Column,
30 m x 0.25 mm (ID), column temperature 2 min at 60 °C, then raised to
220 °C at 4 °C/min and held 7 min. Detection and injection temperature:
300 °C. Sample: 0.05 µL of new Stoddard solvents; split ratio: 200:1.

Figure 8 shows a chromatogram of spent SS. The chromatograms of new and spent
solvent show no significant differences either in terms of new peaks or in concentrations
of existing peaks.

The time-study samples also were analyzed by GC to obtain information on contaminants. Hexane was used as an internal standard to determine the concentrations of
n-alkanes in the time-series and new solvent samples. There was a significant difference
in n-alkane concentrations between the two sets of samples: the concentrations of
decane and tridecane increased by 20 percent and dodecane concentrations decreased by
20 percent in the time-study samples compared with those in the new SS. The concentration of undecane remained practically unchanged.

The change in concentration can be attributed to the common practice of adding
lubricating oil to new SS. The hydrocarbon composition of a typical crankcase lubricant
consists mainly of saturated compounds having roughly a 2:1 ratio between naphthenes
and paraffins (linear and branched-chain).[35] The paraffins present constitute about 25
percent by weight of the oil and consist of low-volatile compounds (C_{11}+). A lubricating
oil also normally contains an additive package that may comprise 20 percent of the oil by
weight.

There was no significant change in n-alkane concentrations within the time-study
samples analyzed, i.e., between the first day's sample (new Stoddard plus oil) and the
other eight samples.

---

[35]*Disposal/Recycle Management System Development for Air Force Petroleum Oils and
Lubricants.*

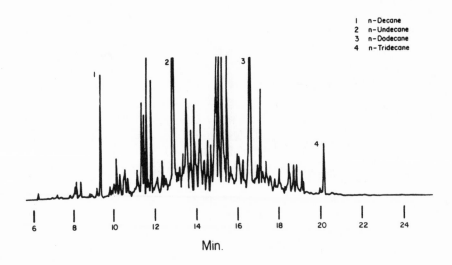

Figure 8.   Spent Stoddard solvent chromatogram. DB-5 Fused Silica Capillary Column, 30 m x 0.25 mm (ID), column temperature 2 min at 60 °C, then raised to 220 °C at 4 °C/min and held 7 min. Detection and injection temperature: 300 °C. Sample: 0.05 μL of spent Stoddard solvent (SS#1); split ratio: 200:1.

The n-alkanes and their respective isomers are responsible for the cleaning action of SS; hence, their apparent nondepletion in the cleaning operation indicates that spent SS could be restored to the specifications of new solvent by simple distillation, without requiring further reprocessing or makeup of lost crucial cleaning component.

Figure 9 shows the variation in solvent KBV for the simulated samples (test series 1). The KBV decreases linearly as the solvent becomes more contaminated. However, the magnitude of the decrease is not great enough to cause a significant decline in solvent power on the overall KBV scale. This finding indicates that the spent solvent is still potent provided it is decolorized. The ASTM standard on new SS specifies a KBV between 29 and 45. The new solvent from Anniston Army Depot has a KBV of 27.5, which is slightly lower than the minimum allowable value specified by ASTM.

Figure 10 shows the variation in KBV of the solvent for the time-study samples. The KBVs show no decrease for the first three samples that cover a 13-day period (one intermediate sample was not measured), then decrease by 0.8 percent and hold steady for the three subsequent samples (9 days), and finally decrease by another 0.8 percent to a value of 26.3 when the solvent was changed. Overall, the decrease in KBV is about 1.5 percent within the time period for which the samples were taken (22 days). The KBV for the first sample (day zero in Figure 10) is 2.9 percent below that of the new solvent due to the addition of lubricating oil. This factor accounts for 66 percent of the decrease that was observed in KBV over the total 22-day sampling period.

The relatively small overall change observed in KBV for the two sets of samples leads to two apparent conclusions:

1. The test is not sensitive enough to be effective for predicting solvent change. However, if a recycling facility exists, this test may serve as a quality control criterion for the reclaimed solvent.

2. The spent solvent does not lose its cleaning ability significantly, and removal of color and particulates will essentially rejuvenate it.

Viscosity

The variation in SS viscosity with increasing contamination (test series 1) is shown in Figure 11. The viscosity of the spent solvent increases by 37 percent over that of new solvent. The increase in viscosity varies quadratically with the percentage of spent solvent added (Figure 11). The significant change is not surprising because it is well known that intermolecular interactions between solute (i.e., the contaminants) and solvent increase as more contaminants are dissolved in the solvent, causing an increase in solvent viscosity.

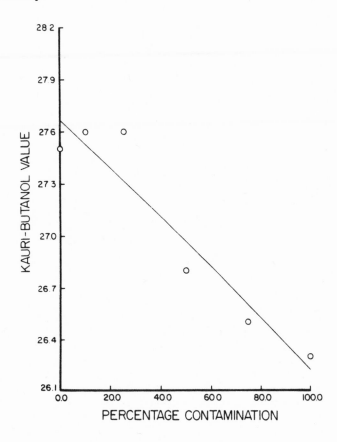

Figure 9. Variation in Kauri-butanol value of Stoddard solvent with contamination.

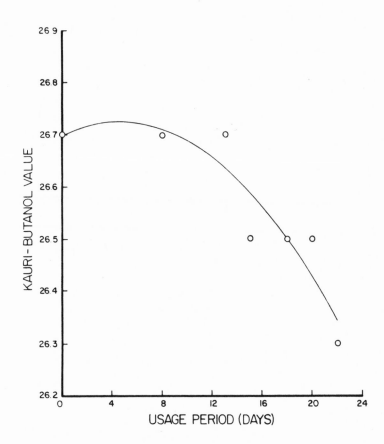

Figure 10. Variation in Kauri-butanol value of Stoddard solvent with usage time.

Figure 12 shows the viscosity variation of SS in response to increasing usage (time-series samples). The first sample shows an increase of about 20 percent over that of the new solvent (see Figure 11), primarily due to the addition of motor oil. The range of increase in viscosity is 14 percent for the entire usage period (22 days), which amounted to an increase of more than 35 percent over that of new solvent. As observed for the KBV measurements, the initial addition of motor oil contributes to more than 60 percent of the total increase observed in viscosity above that of the new solvent.

On the basis of these results, viscosity appears to be a fairly sensitive property. An important factor that must be taken into consideration while measuring viscosity is the effect of temperature. Care was taken so that all measurements were made at the same temperature (usually 20 °C). A small decrease in ambient temperature may cause a significant increase in viscosity and yield erroneous conclusions about the solvent condition. This situation can be remedied in one of two ways: (1) provide a constant temperature bath for immersion of the viscometer or (2) construct several nomograms for viscosity so that the measurements can be corrected to the same temperature. Another factor that must be considered in interpreting changes in a solvent's viscosity is the amount and size of particulates in spent solvent. The presence of particles larger than the diameter of the capillary in an Ostwald viscometer will interfere with liquid flow

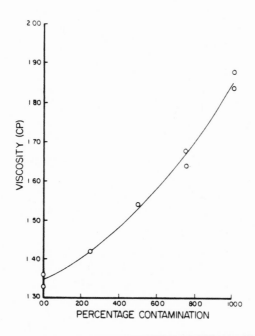

Figure 11. Variation in viscosity of Stoddard solvent with contamination.

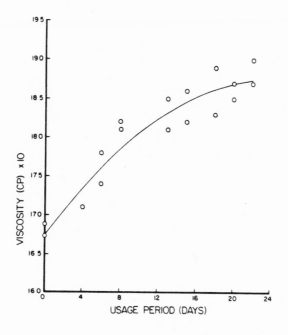

Figure 12. Variation in viscosity of Stoddard solvent with usage time.

through the capillary. For this reason, a spent solvent may have to be filtered prior to measuring its viscosity.

Refractive Index

Refractive index was measured using a Bausch and Lomb (Model ABBE-56) refractometer.[36] The refractive index of water was measured as 1.3309 at 30 °C, which compares with a literature value of 1.3319.[37]

Figure 13 shows the variation in refractive index of SS with increasing contamination (test series 1). The measurements were made at two temperatures--22 and 31 °C. The two isotherms indicate that refractive index increases almost linearly with increasing contamination and decreases with an increase in solvent temperature.

The refractive index profile of SS (Figure 14) with usage time for the time-series samples (test series 2) indicates a fairly significant increase. However, refractive index measurements of contaminated halogenated solvents did not yield the degree of sensitivity seen in past studies with SS at Auburn University. The sensitivity of a solvent's refractive index thus may be contaminant-specific or solvent-specific. As a result, refractive index as an indicator of solvent contamination is promising for monitoring the SS contamination, but it may not be very effective with other types of solvents.

Specific Gravity

The variation in specific gravity for test series 1 is shown in Figure 15 as a function of volumetric percentage addition of contaminated solvent. The presence of dirt and other impurities generally increases the specific gravity of SS. The range of increase seen was about 1.7 percent.

Specific gravity was measured by filling a pycnometer with solvent, followed by weighing on an analytical balance and then repeating the procedure with water. A specific gravity meter was also used to compare and supplement the pycnometer procedure.

Figures 16 and 17, respectively, show the results obtained using a pycnometer and specific gravity meter. The reproducibility of the specific gravity meter measurements was within 0.001, whereas that using a pycnometer and analytical balance was around 0.002. The solvent in the pycnometer evaporates to some extent, thus contributing to a lesser accuracy and reproducibility than can be observed with the specific gravity meter. The specific gravity meter contains an oscillator which is filled with sample and then excited electromagnetically. The period of oscillation is a function of sample density. The density reading is automatically compared with the density of water (a value stored in a microprocessor within the meter), thus producing the specific gravity of the sample. In addition, the densities are evaluated at a set temperature by the meter--usually 20 °C --thus circumventing the need for a temperature control system.[38]

---

[36] *Refractometer Manual*; G. J. Shugar, et al.
[37] *Handbook of Chemistry and Physics*, 57th ed. (CRC Press, 1976).
[38] *Mettler/Paar DMA 35 Density Meter.*

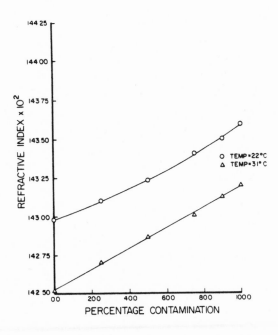

Figure 13. Variation in refractive index of Stoddard solvent with contamination.

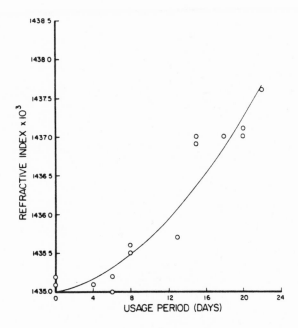

Figure 14. Variation in refractive index of Stoddard solvent with usage time.

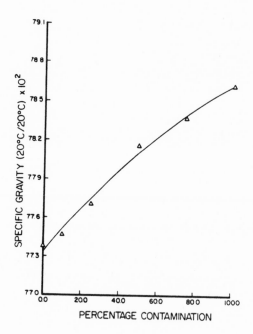

Figure 15. Variation in specific gravity of Stoddard solvent with contamination.

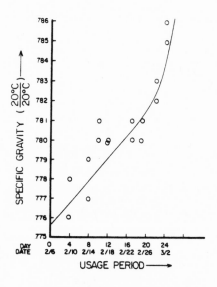

Figure 16.   Variation in specific gravity of Stoddard solvent with usage time using a pycnometer.

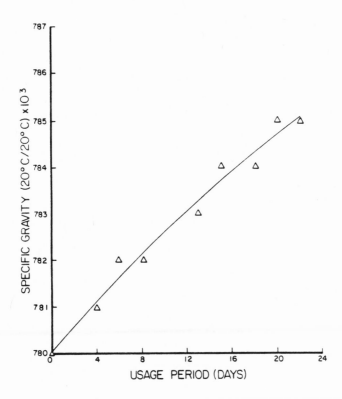

Figure 17.   Variation in specific gravity of Stoddard solvent with usage time using a specific gravity meter.

Apparently, specific gravity is not as sensitive an indicator of solvent quality as is viscosity. However, using a specific gravity meter as a measuring tool makes specific gravity a quick, fairly reliable way to monitor solvent quality.

Electrical Conductivity

Figure 18 shows electrical conductivity measurements of SS as a function of simulated contamination (test series 1). The test is easy to conduct and shows good sensitivity to contamination. The conductivity of spent SS decreases by 17.3 percent with respect to new solvent.

Figure 19 shows the conductivity profile of the time-series (test series 2) samples. In these results, the conductivity also decreases with usage time, i.e., with increasing contamination. The addition of motor oil significantly changes the conductivity of the first sample of the time-series samples (day zero in Figure 19) compared with that of new solvent (35 percent). Thereafter, the variation between the first sample and the final sample of the time-study samples is only about 5 percent. This finding shows that the addition of motor oil has a major effect on this property.

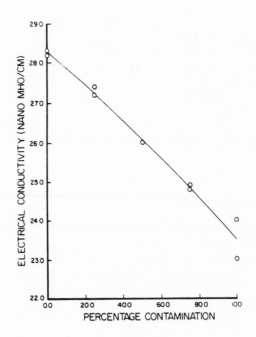

Figure 18. Variation in electrical conductivity of Stoddard solvent with contamination.

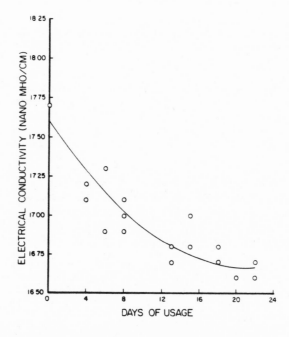

Figure 19. Variation in electrical conductivity of Stoddard solvent with usage time.

The effect of contamination was also studied by adding motor lubricating oil to new SS in different proportions. Figure 20 shows the variation in conductivity of SS after addition of oil. The conductivity initially decreases sharply with increasing oil concentration until about a 20 percent (v/v) oil concentration is reached. The conductivity keeps decreasing at a much slower rate until the curve shows a minimum at roughly a 50 percent (v/v) oil concentration. It then increases with further oil additions for concentrations of oil above 50 percent. As the concentration of oil approaches 100 percent (pure motor oil), the conductivity value approaches that of pure motor oil (which is slightly higher than the conductivity value for new SS). Figure 20 indicates that the conductivity variation for SS is more pronounced at low concentrations of contaminant, and is relatively insensitive at higher contaminant concentrations. It should be mentioned that the high concentrations of oil which cause a reversal in the conductivity slope (Figure 20) are never reached in practical cleaning operations.

Conductivity is a simple, quick test to measure solvent quality. The sensitivity of this test to solvent contamination is also very significant. The only disadvantage of this property as an indicator of solvent quality is the extra care required to clean and maintain the conductivity probe in good condition. The probe may require replatinizing annually if used extensively. This test also requires some form of temperature control because conductivity is highly dependent on temperature.

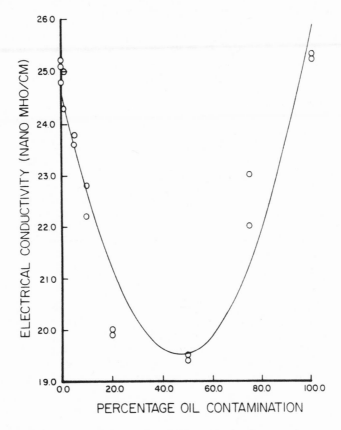

Figure 20. Variation in electrical conductivity of Stoddard solvent with oil contamination.

Visible Absorbence Spectroscopy

The new and spent solvents were first analyzed over almost the entire range of ultraviolet (UV) and visible radiation wavelength spectra (180 to 700 nm). A major peak was observed at 300 nm. However, most spectroscopy studies on SS have been performed between 400 and 600 nm.[39] Accordingly, the investigation was limited to this range of visible radiation.

The visible absorbences of new and various proportions of spent solvent (test series 1) are shown in Figures 21 and 22. The absorbences of new and mixtures containing 25, 50, 75, and 100 percent spent solvent were measured at 400, 450, 500, and 600 nm. In Figure 21, the absorbence was measured at a constant slit width for each of the four wavelengths (the slit width was set to a value at which the new solvent showed zero absorbence for each wavelength). All four curves in Figure 21 followed Beer's law except for a small positive deviation at higher concentrations of contaminants (>75 percent). This deviation from Beer's law may be due to intermolecular attraction between the solvent and its contaminants. In general, solvent absorbence increases with contamination and, as the wavelength decreases, the sensitivity (slope) of the plot increases.

Figure 22 shows the absorbence of constant composition spent solvent at different wavelengths. The variation in absorbence with the addition of various proportions of spent solvent is significant. Typically, a new solvent has a transmittance ranging from 91 percent at 450 nm to 99.1 percent at 600 nm.[40] In the spectroscopic analyses, new solvent had a transmittance ranging from 80 percent at 500 nm to 90 percent between 500 and 600 nm. The spent solvent shows a transmittance of 0.1 percent at 450 nm and about 35 percent at 600 nm. Absorbences at other concentrations between new and spent solvent are also shown in Figure 22; the increase observed in absorbence with increasing contamination is significant.

The time-study samples (time series 2) were also analyzed, and the absorbence profile is illustrated in Figure 23. The discoloration with use, as reflected by increased absorbence, increases almost linearly with usage time. The sensitivity of visible spectroscopy of SS increases with decreasing wavelength.

Visible light absorbence appears to be a reliable, consistent indicator of solvent performance. Absorbence can be measured quickly, and the necessary analytical instrumentation is fairly rugged and requires no special maintenance. For example, a dipping optical probe colorimeter is available that can be dipped directly into a vat or container to obtain an instantaneous readout.

Thin-Layer Chromatography (TLC)

Various dyes were tested to determine those most effective and sensitive for measuring the solvent power of SS. Brilliant Oil Blue BMS (Fischer) and Disperse Yellow 9 (Aldrich cat# 21225-3) worked the best on silica-impregnated glass microfiber sheets. The R_f values of these two dyes are shown in Table 5 for various solvents. Brilliant Oil Blue BMS separates into at least four color components; of these, there are fast-moving violet and blue-1 colors that can be used to characterize solvents of low aromaticity.

---

[39]K. Johnson; I. Mellan; *International Fabricare Institute Bulletin T-447*; *National Institute of Drycleaning Bulletin Service*, T-413.
[40]K. Johnson.

Disperse Yellow 9 adsorbs strongly onto the microfiber sheet, and only solvents with a high aromatic content can transport this dye. Benzene and its mixtures with new and spent SS proved to be very effective in moving both dyes. Aliphatic solvents, such as pure heptane, had no effect on the dyes, and their low KBVs verify that they are poor solvents.

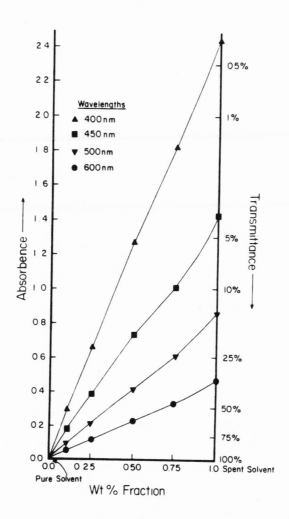

Figure 21. Absorbence of contaminated Stoddard solvent as a function of wavelength.

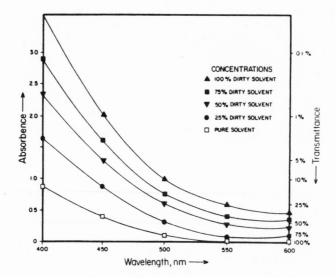

Figure 22.   Absorbence of contaminated Stoddard solvent as a function of volumetric concentration.

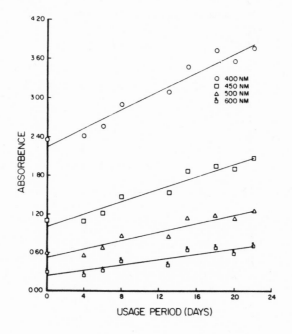

Figure 23.   Variation in visible absorbence of Stoddard solvent with usage time.

Table 5

R_f Values of Different Dyes on Microfiber Paper

	Brilliant Oil Blue BMS				Disperse	
Solvent	Violet	Blue-1	Blue-2	Blue-3	Yellow-9	K-B*
Heptane	≈0.0	≈0.0	≈0.0	≈0.0	≈0.0	17.1
Benzene	0.96	0.84	0.66	0.46	0.53	107.0
25% New Stoddard Solvent 75% Benzene (% Vol.)	0.93	0.73	0.56	0.30	0.23	87.2
25% Spent Stoddard Solvent, 75% Benzene (% Vol.)	0.90	0.71	0.51	0.24	0.23	87.2
New Stoddard Solvent	0.11	0.05	0.04	**	≈0.0	27.6
Spent Stoddard Solvent	0.30	0.22	0.12	0.06	0.06	26.3
Spent Stoddard Solvent Distillate	0.13	0.09	0.06	**	≈0.0	27.7
Spent Stoddard Solvent Bottoms	***	0.44	0.27	0.16	0.11	-

*Some K-B values are from literature; mixture K-B values were computed in proportion to the volume fraction of the pure components.
**This color did not separate distinctly from the dye source spot.
***The color of the solvent overshadowed the faint violet color of the dye.

Both dyes yielded better R_f values with spent SS than with new SS. This finding implies that the new SS becomes contaminated with aromatics during the cleaning operation. To determine the nature of the contaminating aromatics, the spent solvent was distilled and TLC analyses were conducted separately on the distillate and on the bottoms. The R_f values of the dyes with the distillate were identical to those of new SS. The R_f results for the bottoms were significantly better than those with distillate, new SS, and spent SS. This result indicates that the contaminants are heavy aromatics and cannot be distilled easily. This property is beneficial from the standpoint of solvent recycling.

Table 6 shows the R_f value of the dyes for the first and final samples of the time-study series. The final sample shows a small increase in the R_f values for the violet, blue-1, and blue-2 segments of the blue dye relative to the first sample. The R_f value of the yellow dye, however, is higher for the final sample by 40 percent over that of the first sample. This finding suggests that the contaminants contain aromatics, since the yellow dye moves well only in aromatics. However, the KBV for this sample shows no increase above that of the new solvent, which implies that the concentration of aromatic contaminants is very small.

R_f value determination requires an eye estimate of the height the dye travels. This method may lead to erroneous conclusions in cases for which there are minor differences in samples' solvent power. Overall, TLC provides an effective way to compare solvents with significant differences in solvent power, and also to roughly indicate the degree of solvent aromaticity.

Tables 5 and 6 show the KBVs for various solvents. There is a good correlation between R_f values and KBVs. The concept of applying TLC in quantitative determination of solvent power is simple and innovative, and can be used to supplement an established method such as the KBV test.

Rating the Test Methods

Results of the property tests were analyzed and quantified to obtain a relative rating of the methods. Table 7 shows how the test methods were rated. The criteria used for the rating were: sensitivity and reliability; reproducibility; ease of operation; operator training requirements; equipment costs, and maintenance cost.

Table 6

R_f Values of Different Dyes in the Time-Study Series

Solvent	Brilliant Oil Blue BMS				Disperse Yellow-9	K-B
	Violet	Blue-1	Blue-2	Blue-3		
SS #2 (2/6/86)	0.40	0.27	0.18	0.09	0.06	26.7
SS #2 (2/28/86)	0.42	0.32	0.19	0.09	0.10	26.3

Table 7

Rating of Test Methods

Method	Column 2 Sensitivity and Reliability A*(=4, Maximum) B (=3, Good) C (=2, Fair) D (=1, Minimum)	Column 3 Reproducibility A*(=4, Excellent) through D (=1, Poor)	Column 4 Ease of Operation A*(=4, Excellent) through D (=1, Poor)	Column 5 Operator Training A*(=4, Minimum) through D (=1, Extensive)	Column 6 Equipment Cost A*<$500 $ 500<B<$1200 $1200<C<$2000	Column 7 Maintenance Cost (Annual) A*<$50 $200<B>50 $500<C	Rating**
Visible Absorbence	A	A	A	A	C	A	3.8
Density	B	A	A	A	B	A	3.5
Viscosity	B	A	B	A	A	A	3.5
Electrical Conductivity	B	A	A	A	B	B	3.4
Refractive Index	C	A	A	A	C	A	3.0
Kauri-Butanol Value Test	C	A	B	B	A	B	2.9
Thin-Layer Chromatography	C	B	B	B	A	B	2.7

*A=4, B=3, C=2, D=1 Point.
**Rating = 1/10 (4 x Column 2 + 2 x Column 3 + Column 4 + Column 5 + Column 6 + Column 7).

Sensitivity and Reliability

These criteria evaluate the magnitude of change in a property with different levels of solvent contamination. The sensitivity of a method can be correlated as:

$$\%S = [|P_i - P_t| / P_i] \times 100 \qquad [Eq\ 6]$$

where %S is percentage sensitivity, P_i is the property value of the first sample (test series 2), and P_t is the property of the "t"th sample.

Figure 24 is a plot of percentage sensitivity versus usage days for the time-study samples. Visible absorption has a sensitivity of 60 percent between the first and final samples. It is followed by viscosity, conductivity, and density with percentage sensitivities of 13.5, 6.2, and 1.2, respectively. KBV and refractive index methods have relatively lower percentage sensitivities of 0.2 and 0.1, respectively. TLC is not shown in Figure 24 because of the uncertainty involved in measuring R_f values. The levels of grading for each criterion rated were A, B, C, and D, or numerically, 4, 3, 2, and 1 points, respectively.

On the basis of sensitivity studies, visible absorbence was assigned an A ranking, followed by a B ranking for viscosity, conductivity, and density, and a C for KBV and refractive index.

Reproducibility

This factor accounts for the consistency of a method to repeatedly report the same values of a property when tested more than once. All methods showed good reproducibility.

Ease of Operation

This criterion rates the tests according to convenience, that is, the time required to prepare reagents (if necessary) and perform the tests. Visible absorbence, refractive index, specific gravity (by specific gravity meter), and conductivity measurements can be performed rapidly and no reagent preparations are required. One drawback of the viscosity test is that the sample may have to be filtered prior to testing because large particles suspended in the solvent could clog the viscometer's capillary and impede the flow of solvent. KBV measurement requires preparation and weighing of Kauri-butanol solution followed by titration with SS. To perform TLC using glass microfiber paper, a preheating stage is required, and dye solutions have to be prepared prior to use. Consequently, viscosity, KBV, and TLC were ranked lower than the other tests for this criterion.

Operator Training

This criterion rates the preparation and training required by the operator to perform the tests satisfactorily. All test methods in this study are fairly simple and do not require substantial operator training. However, KBV and TLC require relatively more preparation and training as well as operator judgment.

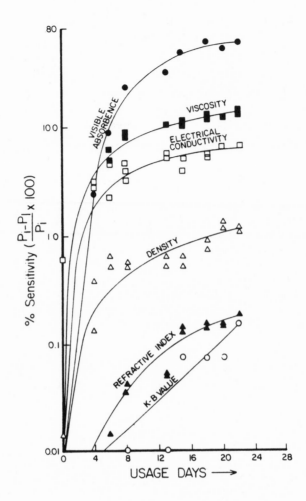

Figure 24. Sensitivity of measurement techniques.

Equipment Cost

This factor is based on the capital cost of the apparatus required for performing each test.   Tests requiring equipment that costs $500 or less were ranked A, those costing higher than $500 but less than $1200 were ranked B, and those with a cost higher than $1200 but less than $2000 were ranked C.   The costs were obtained either from equipment suppliers' catalogs or direct vendor quotes.[41]

---

[41]*Fisher Scientific Catalog* (1986); *Thomas Scientific Catalog* (1986-87); Brinkmann Instruments Co.

The viscosity apparatus consists of an Ostwald viscometer and a stopwatch; the KBV apparatus is basically a buret and an Erlenmeyer flask. These setups cost less than $100 each and were ranked A.

A specific gravity meter and an electrical conductivity meter with the probe cost around $1200 and $1000, respectively, and were ranked B in this category. Refractometers and spectrometers/in-situ dipping probe colorimeters cost about $2000 each; these tests were ranked C.

Maintenance Cost

An annual operating cost was estimated for each test method, taking into account the cost of chemicals, probes, and/or supplies necessary to conduct a test at least once a week. Maintenance cost for test equipment such as the viscometer, refractometer, spectrometer, and specific gravity meter involve buying routine cleaning supplies, and should not exceed $50. Consequently, these methods were ranked A.

In contrast, KBV and conductivity methods as well as TLC analysis require chemical and material supplies costing about $300 annually for each, in addition to routine cleaning supplies. The KBV measurement requires periodic supplies of Kauri-butanol solution. TLC analysis requires glass microfiber paper, dyes, and chloroform. Conductivity probes may have to be replaced annually or require a platinizing kit to replenish the platinum in a used probe.

The overall rating was determined by a formula shown in Table 7. Since sensitivity and reproducibility of a test are of prime importance, they were assigned weights of 4 and 2, respectively. Visible absorbence was rated as the best method, followed by density, viscosity, and electrical conductivity. Refractive index, KBV, and TLC were useful in specific conditions but did not satisfy all criteria of the rating.

Evaluating Batches of Spent Solvent

An important aspect of this study was to establish consistency of the test methods for random batches of spent solvent. To achieve this end, four batches of spent solvent, taken at different times, were obtained from Anniston Army Depot. The first sample (SS#1) was received in October 1985. The second set of samples was obtained in January 1986 and consisted of two samples (TSS#2 and BSS#2) taken from the same vat and at the same time but at different depths. TSS#2 was skimmed off the top of the vat, whereas BSS#2 was taken from the bottom of the vat. The time-series samples (SS#3) were taken in February 1986.

Table 8 shows the variation in physicochemical properties between TSS#2 and BSS#2 samples. The BSS#2 sample contained more particulate matter than did TSS#2. Hence, a significant variation in properties of the two samples would indicate that the amount of particulates affects the properties substantially. However, no significant variation was observed in physicochemical properties except in the absorbences at 500 and 600 nm. The significant variations in absorbences at these two wavelengths could mean that there was some stratification within the contaminated solvent. Occasional stirring may eliminate this problem.

The maximum variation in physicochemical properties among the three spent SS samples is shown in Table 9, and indicates good consistency among the three samples.

Table 8

Physicochemical Tests on Spent Stoddard Solvent Batch 2

Test	Top Spent Batch 2 (TSS#2)	Bottom Spent Batch 2 (BSS#2)	Variation in Batch 2 (%)
Refractive index (22°C)	1.4373	1.4375	0.014
Electrical conductivity, micromho (23°C)	0.0230	0.0230	0.000
Specific gravity (17°C/17°C)	0.7860	0.7870	0.127
Viscosity, cp (18°C)	2.1000	2.2200	5.410
Kauri-butanol value	25.200	25.400	0.790
Visible absorbence:			
400 nm	4.000	4.000	0.000
500 nm	1.850	3.050	39.30
600 nm	1.210	1.540	21.40

Table 9

Physicochemical Tests on New and Spent Stoddard Solvent

Test	New	Spent Batch 1 (SS#1)	Spent Batch 2 (SS#2)	Spent Batch 3 (SS#3)	Maximum Variation (%)
Refractive index (22°C)	1.4298	1.4360 (31°C)	1.4374	1.4376	0.11
Electrical conductivity, micromho/cm (23°C)	0.028	0.024	0.023	0.017	29.20
Specific gravity (17°C/17°C)	0.774	0.785	0.787	0.785	0.25
Viscosity, cp (18°C)	1.350	2.200	2.160	1.880	14.50
Kauri-butanol value	27.600	25.300	25.300	26.500	4.50
Visible absorbence:					
400 nm	1.010	3.370	4.000	3.770	15.80
500 nm	0.170	0.950	2.450	1.230	61.30
600 nm	0.000	0.440	0.380	0.690	44.90

The value shown for SS#2 is the average of experimental values for TSS#2 and BSS#2. The conductivity of the spent solvents showed the greatest variation (29.2 percent), whereas specific gravity values were remarkably close (0.25 percent) among the three samples.

The absorbence at 400 nm indicates a difference of 15.8 percent between SS#2 and SS#1 samples. This result may be due to greater particulate matter in SS#2 than SS#1. The observed difference in absorbence becomes insignificant considering the fact that, for absorbences greater than or equal to 3, the transmittance is negligible.

SS#3 had significantly lower viscosity and conductivity than the other spent samples. This lower viscosity indicates that it was less contaminated than the other samples. On the other hand, the lower conductivity would signify that it was more contaminated than the other samples. These deductions show that relying on just one test method may allow misleading conclusions, resulting in premature removal of solvent or, in some cases, unsatisfactory cleaning.

SS is a mixture of many compounds and there is substantial variation in the properties of new solvent from batch to batch. Thus, it is best to convert the property measurements into nondimensional quantities so that the rating criteria are independent of batch variations.

Table 10 shows the reconstruction of Table 9 into respective nondimensional properties for some of the test methods. The nondimensional quantities for the spent solvent are obtained by the following relationship:

$$NDP = |\ C_m - C_o\ |\ /\ s \qquad\qquad [Eq\ 7]$$

where NDP is the nondimensional property under study, C_m is the measured property value of spent solvent, C_o is the measured property value of new solvent, and s is the estimated measurement error of a test method.

This relationship of NDP is arbitrary and can be changed to fit the requirements of a particular situation. For example, C_o could be substituted for s to obtain NDP. This NDP, as defined above, conceptually represents the magnitude of change in a measured parameter from the new solvent value to the spent solvent value relative to the measurement error of a method. (The average of the three solvent batches also is shown in Table 10). If the average NDP is taken as the cutoff point to change a solvent, it can be observed that only SS#2 satisfied the cutoff points of visible absorbence, viscosity, and specific gravity. The electrical conductivity cutoff point was exceeded only by the SS#3 sample. Sample SS#1 did not meet the cutoff limits of any of the tests.

Discussion of Findings

From the results of the physicochemical test methods, the following conclusions are proposed:

1. At present, DOD uses inadequate, empirical field tests for solvent management. No standard test(s) is being used for predicting the need to change solvent. Development of a test(s) will lead to optimal solvent usage without compromising cleaning performance.

Table 10

Nondimensional Property Data on Spent Stoddard Solvent

Test	Estimated Measurement Error	Spent Batch 1 (SS#1)	Spent Batch 2 (SS#2)	Spent Batch 3 (SS#3)	Average
Electrical conductivity (nanomho/cm)	0.300	13.3	16.0	36.7	22.0
Specific gravity	0.001	11.0	12.5	11.0	11.5
Viscosity (cp)	0.020	42.5	40.5	26.5	36.5
Visible absorbence: 400 nm	0.010	235.2	299.0	275.2	269.8

2. Laboratory tests show that some physical, chemical, and electrical properties of SS can be used successfully as criteria to identify spent solvents. Visible absorbence was rated as the best criterion among the methods evaluated. Specific gravity, viscosity, and electrical conductivity were also determined to be effective criteria. Kauri-butanol value, TLC, and refractive index methods did not satisfy all of the requirements for reliable criteria, but they are useful in specific situations.

3. These criteria were tested over several batches of spent solvent and were found to be consistent. However, the identification of spent solvent is most accurate when at least two or more criteria are used; reliance on a single criterion may occasionally lead to erroneous conclusions, resulting in either premature removal of solvent or unsatisfactory cleaning performance through prolonged use.

4. Reclamation and Reprocessing Techniques

Most solvents identified in large quantity in waste streams at military installations can be recycled using reclamation equipment, e.g., filtration or distillation units. Reclamation by filtration is mostly limited to transmitter and electrical equipment coolants (e.g., FC-77, DC-200, and 25R-F15 coolants). The filters consist of 0.1-micron synthetic fibers.[42]

The other recycling technique, which is the most prevalent, is distillation to separate volatile solvent materials from less volatile contaminants. There are two major distillation methods used in reclaiming solvents: simple flash distillation and fractional distillation. The particular method used depends on the types and amounts of solvent and contaminants, and the specifications set for the reclaimed solvent. In most situations, solvent separation can be effected by simple distillation, with no enrichment trays or packing required. Proper segregation of the various solvents is necessary to reclaim the material by simple flash distillation. Fractional distillation in which the column has an enrichment tray or packing section is used when either solvent purity specifications are stringent or segregation of used solvents is not practiced properly.

This study is limited to a general discussion of simple distillation reclamation, with a more detailed look at batch distillation. DOD has identified batch distillation as one of the most promising processes for solvent recycling on military installations.

Methodology

A pure liquid boils when its vapor pressure equals the ambient pressure. In the case of a mixture, the total mixture vapor pressure must equal the ambient pressure to allow boiling. A pure component boils at a specific temperature, at a given pressure, known as the boiling point (BP). For example, at 1 atm pressure, water boils at 100 °C. A multi-component mixture boils within a temperature range, depending on its composition.

A liquid's BP decreases as the applied pressure decreases. Figure 25 shows the behavior of several halogenated solvents. As an example, methylene chloride boils at 40 °C at atmospheric pressure (760 mm mercury). However, if the ambient pressure is reduced to 400 mm of mercury by applying vacuum, the boiling point of methylene chloride is lowered to 25 °C. This property can be advantageous for high-BP solvents (e.g., SS) if the solvent decomposes, flashes, or ignites above a certain temperature. By knowing the solvent's flashpoint, distillation can be performed at a temperature lower than that value, which ensures safety. The BP can be lowered by using vacuum or by adding an azeotropic component such as water. When water is added to an immiscible solvent, two distinct liquid phases form. The added vapor pressure of water causes the liquid to boil at a lower temperature than if only solvent were present. The significance of codistillation with water or steam is that the solvent can be reclaimed at a temperature lower than its normal BP as illustrated in Table 11. The most common method of codistillation is to add live steam to the charge in the distillation chamber.

The three basic functions of batch distillation are evaporation, enrichment, and collection. Figure 26 outlines these three functions on a typical commercial distillation unit.

---

[42]J. B. Systems, LaGrange, GA.

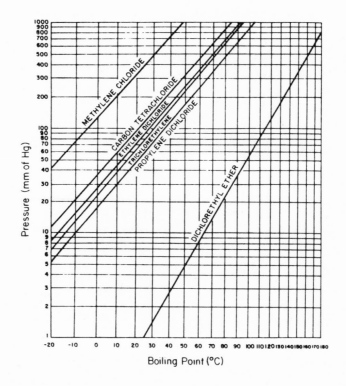

Figure 25.   Boiling point variation with pressure for various chlorinated solvents. (Source: I. Mellan, *Industrial Solvents*, 2nd ed. [Reinhold, 1950]. Used with permission.)

In simple batch distillation, spent solvent is first charged to a vessel. The liquid charge is boiled and the vapors are condensed and collected as purified solvent. The contaminants are left behind in the vessel.

The particular unit shown in Figure 26 uses live steam injection to evaporate the solvent. Another codistillation method is to use an electric heat source and add water to the charge. The liquid distillate is pumped to a gravity separator to remove the water, and the clean solvent is sent to storage.

Evaporation

The important factors in the evaporation stage of distillation are:

- The rate of evaporation is determined primarily by the rate at which heat is supplied to the distillation vessel.

- At very high temperatures, the solvent could degrade. Reducing the pressure within the system lowers the boiling point. Also, equipment limitations may prevent the use of extremely high temperatures.

Table 11

Comparison Between Normal Boiling Points of Different Solvents and Codistillation Boiling Point With Water*

Solvent	Boiling Point (°F)		Solvent Phase: Water Phase Ratio
	Atmospheric	Codistillation	
Hexane	157	142.9	25.3:1
Heptane	209	174.8	9.87:1
Stoddard	308-316	204	0.9:1
Benzene	176	157	10.24:1
Toluene	232	185	3.95:1
Xylene	261-318	202.1	1.5:1
Trichloro-ethylene	189	163.8	9:1
Tetrachloro-ethylene	249	189.7	3.4:1
1,1,1-Tri-chloroethane	166	149	15.8:1
Methylene chloride	104	101.2	61.5:1
Freon TF	117.6	112	9.5:1
Freon 112	199	166	8.5:1
MIBK	241	190.2	4.1:1

*Source: DCI International, Indianapolis, IN. Used with permission.

Two ways of lowering the boiling point are: (1) operate under a vacuum, i.e., use vacuum distillation, or (2) use a mass transfer medium. For example:

- Use codistillation with an insoluble lower BP material such as water. In this case, an azeotropic mixture will boil at a lower temperature than pure solvent.

- Pass a stripping vapor such as steam through the solvent. The steam heats the solvent, raising the vapor pressure while mass transfer occurs from the solvent to the water vapor phase.

Enrichment

As seen from the typical commercial distillation unit, the enrichment section usually plays a very small part in solvent reclamation. The enrichment section in this commercial distillation unit consists of a mist impinger.

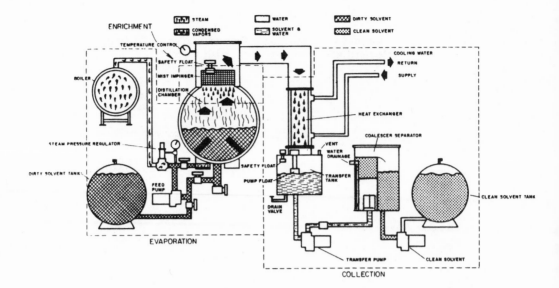

Figure 26. A typical commercial bath distillation unit. (Based on the Dyna I System, Live Steam Injection Model, DCI Corp.)

Condensation and Collection

The condenser is the main piece of equipment within the collection section (see Figure 26). In the condenser, the solvent vapors are cooled either by air, water, or refrigerant, and are condensed to a liquid. It may be necessary to separate the two immiscible liquid phases collected (e.g., water and solvent) by gravity settling and decanting.

Problems in Batch Distillation Operations

Several notable problems are involved in batch distillation. First, thermal degradation can be caused by poor heat transfer within the distillation chamber. The bottom of the chamber is usually hotter than the middle or top section. A very high bottom temperature could cause chemical breakdown of the solvent which may hinder its cleaning performance. Another major problem often encountered in batch distillation is solvent purity. Impurities in the recovered solvent or a loss of inhibitors during solvent recovery could hinder the overall effectiveness of the reclaimed solvent. Inhibitors or new solvent may have to be added to the recovered material to make it effective.

Solvent Characterization

Table 12 classifies solvents in terms of type and major process uses. For typical DOD applications, batch distillation is most attractive when the same unit is used to reclaim multiple solvents. The major solvents (excluding SS) that are recoverable by distillation are described below.

Table 12

Solvent Application on DOD Bases*

Operation Category	Solvent Type
Vapor degreasing	Halogenated 1,1,1-Trichloroethylene Tetrachloroethane Trichloroethylene
Cold-dipping vats	Mineral Spirits Stoddard Solvent PD-680
Paint and carbon removal	Methylene chloride with additives** (phenol, ethanol, petronate HL, water, toluene, paraffin, sodium chromate, methyl cellulose)
Paint thinners	Methyl ethyl ketone Toluene xylene
Metal preparation and precision cleaning	Alcohols Freons

*Source: R. W. Bee and K. E. Kawaoka, *Evaluation of Disposal Concepts for Used Solvents at DOD Bases*, TOR-0083(3786)-01 (The Aerospace Corp., February 1983). Used with permission.
**Additives list was obtained from Robins AFB, GA.

Vapor Degreasers

Vapor degreasing of metal parts is done using chlorinated solvents. Because of EPA regulations, 1,1,1-trichloroethane is gradually replacing trichloroethylene and tetra-chloroethylene at many DOD installations. This compound is less toxic than most halogenated solvents.

Paint Thinners

Paint thinner compounds include toluene, xylene, methyl ethyl ketone, and alcohols. The three major applications of paint thinners are to: (1) thin paint and coatings before application, (2) clean surfaces prior to painting and, (3) clean paint application equipment. The first two areas have wastes from which little or no solvent can be reclaimed; besides, thinners used in these areas require tightly controlled formulation. The third area produces a waste stream that yields a significant amount of solvent(s) which can be reclaimed. Solvent formulation for this application is not very stringent.

Thinners are generally mixtures of several solvents. Thus, it is not practical to use batch distillation to isolate each solvent unless several passes are made.

Paint Strippers and Carbon Removers

Paint strippers and carbon removers consist mainly of methylene chloride mixed with additives. The spent solvent has high disposal costs because it contains phenols and salts. Manufacturers of those strippers generally have a take-back policy because of their reprocessing capability. Material reformulation is required to reuse the solvent as a stripper.

Precision Cleaners

The typical cleaning solvents are ketones and esters, which are used for cleaning surfaces prior to painting. Freons are used to clean electrical parts and appliances, and for leak detection.

Batch Distillation Unit Design

Consultation with various solvent reclaimers and reclamation equipment manufacturers yielded a set of five criteria that should be considered to obtain the properly sized unit for individual operation:

1. Types of solvents to be reclaimed.

2. Types of contaminants in solvent.

3. Amount of solvent.

4. Utilities available.

5. Cycle times of solvents.

Since most installations must recycle more than one solvent, this information must be considered for all solvents. The type of solvent and contaminants must be specified to ensure proper selection of construction materials and waste-handling design.

The utilities available determine the heat source for the evaporation section. Most industrial batch distillation units are manufactured for steam heating but are convertible to natural gas or electricity.

Residence time of the solvents reclaimed is critical to proper sizing of batch units. The manager must consider the time required to reclaim each solvent used as well as the order of reclamation. Table 13 summarizes distillation equipment sizes. For large-scale operations, such as the 50 to 125 gal/hr range, continuous operation is applied.

Batch distillation requires little operator time. Most attention is required at the beginning and end of each cycle, with occasional monitoring during the cycle. Two methods used to determine the end of the cycle are temperature setpoint and cycle time. For a temperature setpoint system, once a certain temperature is reached, the operation is stopped. For the cycle-time method, a set amount of time is allowed to pass and the distillation is terminated.

Table 13

Typical Commercial Solvent Distillation Equipment*

	Size		
Item	Small	Medium	Large
Flow rate (gal/hr)	15	15-50	50-100
Operation type	Batch	Batch or continuous	Continuous
Solvents reclaimed	Paint thinners, chlorinated solvents	All solvents	All solvents
Attention required	Automated shut-down at end of batch	Automated, requires operator attention at end of batch	Automated with occasional operator checks
Cost (1987 dollars)	3,000-5,000	30,000-60,000	60,000-100,000

*Source: R. W. Bee and K. E. Kawaoka, *Evaluation of Disposal Concepts for Used Solvents at DOD Bases*, TOR-0083(3786)-01 (The Aerospace Corp., February 1983). Used with permission.

Solvent distillation equipment ranges in size from small, independent units (0.5 to 15 gal/hr) to large units capable of processing 100 gal/hr or more (Table 13). The small units are generally available in both atmospheric and vacuum distillation modes. These systems are self-contained and require only electrical and cooling water supply connections. The large units require more extensive support facilities (e.g., a boiler). The units are modular and may have an operating lifetime of 20 years or more if maintained properly. The operating costs include labor, utilities, and maintenance material.

Operator time is required only when loading and unloading charges (in the case of batch distillation) and occasional monitoring. Because the types of solvents recycled are relatively constant, operation of the units is fairly easy and requires only a few weeks of operator training.

Utilities expenses consist of electricity and cooling water costs. The electricity cost is a function of solvent type and the degree of contamination in the used material stream, whereas cooling water requirements are consistent for all solvent types.

Table 14 shows distillation equipment costs obtained from two manufacturers. The sizes are based on the amount of product collected per hour. The DCI unit includes the still and separation equipment. When choosing a system, the installation must consider the parameters that will allow maximum flexibility (e.g., Robins AFB has a vacuum system to handle SS even though Freon, another solvent reclaimed, needs no vacuum). To optimize the reclamation of multiple solvents in a single unit, the order of reclamation also must be considered.

If a large inventory is not available, the solvent order may have to be staggered so that all solvents are reclaimed according to their rate of generation. In this case, proper coordination between the waste generation schedule, reclamation schedule, and the usage schedule is essential. Optimization techniques, e.g., linear programming, may be useful in this regard.

Experimental Reclamation of Spent Solvent

Reclamation studies on spent SS consisted of investigating decolorizing and filtration methods along with batch distillation using bench-scale and pilot units. Before proceeding with any one recycling technique, the concentration of chlorides in spent solvent samples was determined. This step is important because, if a substantial amount of chlorides is present in the spent solvent, it may carry over in recycled solvent and could pose potential health hazards to workers upon reuse in open vats.

Chloride Analysis

The ASTM standard test method requires combustion of the sample in a Schoniger combustion flask followed by a potentiometric titration. This method was attempted but found to be time-consuming and relatively insensitive to low chloride concentrations. Combustion of the oil sample in a Parr Bomb, followed by an Argentometric titration, was then tried and found more suitable.[43] A small sample of waste oil in a Parr Bomb with 10 mL of $NaHCO_3/Na_2CO_3$ solution was charged to 24 atm with oxygen and ignited. The bomb was next allowed to sit for 10 min and was then quantitatively rinsed with deionized water to a volume of 100 mL. Potassium chromate indicator solution (1 mL) was then added and the solution adjusted to a pH range of 7 to 10. The solution was then titrated with 0.0141N silver nitrate solution until a distinct color change took place. The results correlate well with the Beilstein test.[44] The chloride test results indicate that samples of spent solvent (SS#1) contain about 1000 ppm of total chloride. This concentration is well below the 4000-ppm level set by EPA for hazardous wastes.

Table 14

Typical Distillation Equipment Costs

Manufacturer	Size (gal/hr)	Cost ($)	Additional Costs for Vacuum Systems ($)
DCI*	100	42,600	7,600
	500	63,100	11,200
	1000	77,800	15,700
Gardner**	50	50,000	Included
	100	70,000	Included
	200	103,000	Included
	500	133,500	Included

*Data obtained from DCI Corp., 5752 W. 79th St., Indianapolis, IN.
**Data obtained from Gardner Machinery Corp., 700 N. Summit, Charlotte, NC.

[43]*Evaluation of Chlorine Determination in Waste Oils*, Contract No. 68-01-7075 (USEPA, 1968).
[44]*Development of a Field Test for Monitoring Organic Halides in Waste Fuels*, Contract No. 68-01-7075 (Auburn University, 1968).

Decolorization and Filtration

An effort was made to isolate the contaminants by contacting the spent solvent with various oxidizing reagents. Sulfuric acid, hydrogen peroxide, and sodium thiosulfate produced a turbid emulsion (Table 15). Sodium hypochlorite and activated charcoal treatments resulted in some improvement in color. A proprietary type of clay[45] produced the most significant improvement in the color of the spent solvent.

Figure 27 shows the effects of the successful reagents in terms of visible absorbence spectroscopic analysis. The curves with the highest and lowest absorbences are those of spent and new SS, respectively. The intermediate curves represent the degree of color improvement by contacting with some of the reagents listed in Table 15. Sodium-hypochlorite-treated solvent was rejuvenated to an absorbence level of 25 percent spent solvent.

Solvent treated with activated charcoal showed little or no change in the 400 to 450 nm range, but improved significantly in the 500 to 600 nm range. Viscosity measurement of the charcoal-treated solvent showed no appreciable difference relative to the spent solvent (1.82 cp). Spent solvent treated with clay showed absorbence characteristics similar to that of new solvent. The solvent treated with clay which was used previously also showed a significant color improvement. Both clay and charcoal adsorbents can be contained in filter cartridges.

An important factor in determining the economic feasibility of clay filtering is how to safely dispose of what could be a voluminous amount of spent clay. Therefore, reclamation through clay filtration versus distillation would depend on the amount of solvent used at the installation.

Table 15

Regeneration Techniques Used To Reclaim Spent Solvent

Sodium Hypochlorite (6%) Reagents	Significant Change Effect
Absorbing Agents:	
Sulfuric acid	Turbid solvent
Hydrogen peroxide (30%)	Turbid solvent
Sodium thiosulfate	Turbid solvent
Sodium hypochlorite (6%)	Color improvement
Adsorbing Agents:	
Activated charcoal	Color improvement
Clay (china clay)	No change
Clay (infusurial earth)	No change
Clay (proprietary)*	Significant change

*Data obtained from J. B. Systems, LaGrange, GA.

---

[45] J. B. Systems, LaGrange, GA.

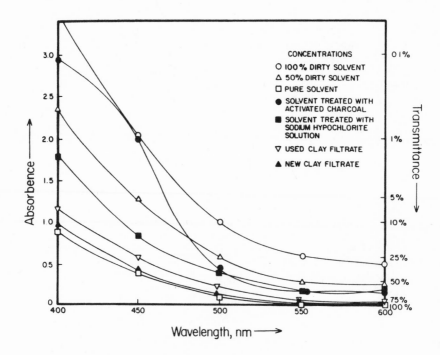

Figure 27. Absorbence of contaminated solvent treated with bleaching agents.

Batch Distillation

Distillation of spent SS was studied experimentally on a bench scale and in pilot units. The boiling range of SS Type II is between 350 and 412 °F. The bench-scale distillation unit was connected to a vacuum pump capable of producing less than 1 mm mercury absolute pressure. This pressure reduced the boiling range to 100 to 200 °F and approximately 90 percent of the solvent was recovered with ease. The reclaimed solvent was colorless and was analyzed using GC. The chromatogram was very similar to that of new solvent, with no significant new peaks or changes in concentration of major compounds detected. The distillation residue, or bottoms, was also analyzed. Many heavy compounds (beginning with dodecane and continuing to those substantially heavier than hexadecane) were observed in minute concentrations. A TLC analysis of the bottoms showed that R_f values of the dyes in bottoms were significantly higher than those with distillate, new, and spent SS (Table 5). This finding indicates that the residue probably contains high-BP aromatics and/or polar species.

The Auburn Waste Oil Reprocessing Laboratory of Auburn University has access to a 15-gal batch distillation unit for fractionation studies. This pilot-scale unit was used to reclaim spent SS. Since the boiling range of SS is between 350 and 412 °F (Type II), water was added to the solvent to generate an azeotropic mixture that has a boiling point substantially lower than that of the solvent alone (see Table 11). The azeotropic mixture containing the solvent and water vapor can be separated from impurities in the mixture, which do not form part of the azeotropic mixture. The water/solvent vapor is condensed and the water layer is separated by gravity.

Approximately 5 gal of spent solvent were charged into the still and 90 percent of the solvent was reclaimed. The reclaimed solvent was tested for quality and its comparison with new solvent is shown in Table 16. The properties evaluated were KBV, refractive index, viscosity, visible absorbence, electrical conductivity, and specific gravity. The properties of spent solvent show a significant difference compared with new and reclaimed SS. The properties of new and reclaimed solvent are remarkably close, thus indicating that the solvent loses no vital cleaning component during either reclamation or previous usage.

For SS, the visible absorbence, viscosity, electrical conductivity, and specific gravity appear to give the best indication of when to remove solvent from a vat, and also can be used to perform quality control checks of recycled solvent. For trichloroethylene and other halogenated solvents, absorbence, electrical conductivity, and acid acceptance value give the best measure of solvent contamination and quality after recycling.

Reclamation Economics

Robins AFB has reclaimed various spent solvents for several years. Table 17 shows the annual savings. These data indicate that Robins AFB is consistently saving up to $0.5 million annually through solvent reclamation.

Table 16

Physicochemical Tests on New, Reclaimed, and Spent Stoddard Solvent Samples

Test	New	Reclaimed by Distillation	Spent
Kauri-butanol value	26.5000	27.1000	25.3000
Refractive index	1.4394	1.4396	1.4375
Viscosity (cp, 20°C)	1.3500	1.3600	1.8000
Visible absorbence:			
400 nm	1.2900	1.3000	3.7900
500 nm	0.2800	0.3100	1.3300
600 nm	0.0300	0.0000	0.9900
Electrical conductivity (nmhos/cm)	22.0000	21.9000	14.8000
Specific gravity (26°C/26°C)	0.7740	0.7680	0.7860

Table 17

Material Savings Trend at Robins AFB by Reclamation*

| Item | Fiscal Year | | | |
	1982	1983	1984	1985
Material reclaimed** (gal)	5060	13,860	20,587	14,418
Material savings ($K)	256.7	404.1	490.3	384.4
Cost avoidance (drums turn-in)	7.6	20.8	30.6	21.6
Gross savings ($K)	264.3	424.9	520.9	406.0
Reclamation costs ($K)	3.3	5.7	7.7	5.9
Total net savings ($K)	261.0	419.2	513.2	400.1

*Source: Robins AFB, Warner-Robins, GA.
**Solvents reclaimed were FC77, Freon 113, trichloroethylene, and isopropyl alcohol.

A literature survey on solvent reclamation at other DOD bases indicates that annual cost savings are already significant at installations that reclaim solvent; thus, the savings could be substantial if this practice were applied to all DOD installations. Table 18 summarizes the potential cost savings (annual) for military bases.[46] Lee et al. have estimated the potential savings if most of the solvents used at NARFs were reclaimed.[47] These savings are shown in Table 19, and range from $0.35/gal for Stoddard to $8.40/gal for Freon TF. The potential savings for the Navy amount to $1.1 million by reclamation of spent solvents at the NARFs.

A rough estimate of distillation costs by Bee and Kawaoka shows that about $0.50/gal of solvent is incurred for a 50-gal/hr distillation unit.[48] The cost of new SS is around $1.80 to $2/gal.[49] Thus, the potential savings is fourfold in the case of Stoddard, and even more for halogenated solvents.

---

[46] R. W. Bee and K. E. Kawaoka.
[47] H. J. Lee, I. H. Curtis, and W. C. Hallow.
[48] R. W. Bee and K. E. Kawaoka.
[49] *Chemical Market Reporter* (October 1986).

Table 18

Annual Cost Savings for Large and Small Bases ($K)*

Solvent	Cost Avoidance per Base		Savings	
	New Material	Disposal	Each Base	All Bases
Large Bases (29)				
Trichloro-ethylene	74.250	36.000	110.250	3200
Stoddard solvent	39.000	9.900	49.500	1400
Paint stripper**	0	40.425	40.425	1200
Paint thinner	41.940	7.555	49.995	1400
Freon	28.512	3.037	31.549	500
Subtotal	183.702	96.917	281.719	7700
Small Bases (124)				
Stoddard solvent	11.880	3.000	14.880	1900
Paint thinner	5.130	1.000	6.130	700
Subtotal	17.010	4.000	21.010	2600

*Source: R. W. Bee and K. E. Kawaoka, *Evaluation of Disposal Concepts for Used Solvents at DOD Bases*, TOR-0083(3786)-01 (The Aerospace Corp., February 1983). Used with permission.
**Manufacturer take-back option.

Table 19

Summary of Reclamation Economics for Solvents Used at NARFs*

Location	Savings in Stoddard ($K)	Savings in Six Most Used Solvents ($K)
Alameda, CA	28.3	192.2
Norfolk, VA	7.5	194.2
North Island, CA	85.5	278.5
Pensacola, FL	18.0	297.5
Jacksonville, MI	2.4	126.7
Total	141.7	1087.0

*Source: H. J. Lee, I. H. Custis, and W. C. Hallow, *A Pollution Abatement Concept, Reclamation of Naval Air Rework Facilities Waste Solvent, Phase I* (Naval Air Development Center, April 1978).

5. Testing and Reclamation Guidelines

In implementing a used solvent management program, the installation DEH must be aware of the solvent types, quality and quantity of waste solvents, segregation practices, current disposal practices, and feasibility of on-/off-site recycling/disposal options. This report has discussed in detail the variety of possible testing procedures to determine the quality of Stoddard-type waste solvents and their reclamation by batch disposal. Two additional documents are available which describe the wide variety of other solvents used at DOD installations and explain management strategies.[50]

Based on the test methods evaluated in this report, some general guidelines can be recommended for testing and reclaiming used solvents. These recommendations are summarized below.

Testing Used Solvents

Subject solvent samples to the following tests:

1. Measure the absorbence with an optical probe colorimeter or a hand-held spectrometer placed directly into a vat or container to obtain an instantaneous readout of visible absorbence (at 500 nm). Assign a rating value (0 to 4) based on Table 20.

2. Measure the specific gravity using an electronic specific gravity meter or pycnometer as described in Chapter 3. Assign a rating (0 to 3) based on Table 20.

3. Measure the viscosity using an Ostwald viscometer according to the procedure described in Chapter 3. Assign a rating (0 to 2) based on Table 20.

4. Measure the electrical conductivity with a conductivity meter and assign a rating (0 or 1) based on Table 20.

5. Sum the individual ratings in steps 1 through 4 above to obtain an overall rating for the solvent.

6. If the overall rating is greater than or equal to 6, the solvent can be considered spent and is a candidate for reclamation. Solvents with a rating of less than 6 should be reused.

The cutoff values presented in Table 20 are suggested based on the results of spent solvent tested in this work. They should be modified as further experience is gained with measurements at an installation.

---

[50]B. A. Donohue and M. B. Carmer, *Solvent "Cradle-to-Grave" Management Guidelines for Use at Army Installations*, Technical Report N-168/ADA137063 (USA-CERL, November 1983); *Used Oil and Solvent Recycling Guide* (Robert H. Salvesan Associates, 1985).

Table 20

Test Criteria for Used Solvents

Rating	Tests			
	Absorbence (500 nm)	Specific Gravity (17 °C)	Viscosity, cp (18 °C)	Conductivity, nmho (23 °C)
0	< 0.6	< 0.773	< 1.35	< 22.5
1	0.6 - 0.8	0.773 - 0.779	1.35 - 1.85	> 22.5
2	0.8 - 1.0	0.779 - 0.785	> 1.85	
3	1.0 - 1.2	> 0.785		
4	> 1.2			

Reclamation

Onsite distillation is considered the most promising process for reclamation of spent solvents. It has been shown to be cost-effective, easy to implement, and easy to operate.[51] The following procedures are recommended:

1. Locate solvents for reclamation.

2. Determine the quality and quantity of waste solvents.

3. Select and install suitable equipment and facilities for reclamation.

4. Provide shop and departmental segregation facilities and mandate handling procedures to assure minimum contamination of wastes with undesirable materials. Color coding may be useful.

5. Place the distillation apparatus into operation.

6. Test the reclaimed solvent according to procedures outlined previously, and categorize and store them for reuse.

---

[51] Higgins (1985).

6. Conclusions and Recommendations

This study is investigating methods of testing used cleaning solvents to determine the point at which they should be reclaimed or discarded. To be successful at military installations, these tests would have to be accurate, consistent, and simple to perform in the field. This work also is examining methods of reclaiming spent solvents in an effort to reduce the amount of this material that must be destroyed. By recycling solvents, installations could save money on new material and disposal charges while ensuring that the amount of potentially hazardous waste solvent entering the environment complies with future regulations, which are expected to be stringent.

Phase I of this work focused on Stoddard solvents. The literature was surveyed to identify properties of these solvents and their current use at DOD facilities. The most common solvents used within DOD are vapor degreasers, cold-dipping cleaners, paint thinners, paint strippers and carbon removers, and precision cleaners.

Another literature survey generated a list of physicochemical tests with potential application at military installations. The methods were assessed, compared, and rated in terms of sensitivity and reliability; reproducibility; equipment cost; and maintenance cost. For most military cleaning operations, the methods showing the most promise in judging solvent quality, in order of decreasing preference, are: visible absorbence, viscosity, conductivity, density, Kauri-butanol value, and refractive index. Besides providing a meter for changing solvents, these tests could be used to monitor the quality of reclaimed product.

Solvent reclamation by filtration and/or distillation was investigated to determine the feasibility of using these techniques at installations. Based on the track records of these methods as reported in the literature, bench- and pilot-scale studies were conducted using spent solvents obtained from Anniston Army Depot. The results indicated that the cleaning properties of these solvents are not lost during either use or reclamation; thus, it would be technically feasible to reclaim these materials throughout DOD. In addition, these methods are economically attractive. Most solvents can be reclaimed using batch distillation, with the most cost-effective system handling distillation of multiple solvents in one unit. Simple units are available which may be able to reclaim up to 98 percent of the solvent. The payback period for such systems can be less than 2 years.

In some cases, filtration to trap particulates from the spent solvent is an effective method of reclamation. The filter material usually is some type of clay. It must be recognized, however, that the process generates waste clay and thus would be practical only for installations using small amounts of solvent.

Based on an evaluation of the testing and reclamation techniques available for Stoddard solvents, guidelines were developed for use in an installation waste solvent management program. A simple numerical rating system has been proposed to allow quick assessment of the solvent and determination of reuse versus recycle.

These findings suggest that substantial savings could be realized if solvents were reclaimed at all DOD installations that conduct regular cleaning operations with these chemicals. In addition, less waste solvent would be targeted for disposal, with the dual effect of lowering disposal/handling costs and resolving environmental issues. On this basis, it is recommended that all DOD installations which generate waste solvent explore the possibility of implementing batch distillation and/or filtration to reclaim spent solvents.

References

Annual Book of ASTM Standards (American Society for Testing and Materials [ASTM], 1983).

Bauer, H. H., G. D. Christian, and J. E. O'Reilly, *Instrumental Analysis* (Allyn and Bacon, 1979).

Bee, R. W., and K. E. Kawaoka, *Evaluation of Disposal Concepts for Used Solvents at DOD Bases*, TOR-0083(3786)-01 (The Aerospace Corp., February 1983).

Bunge, A. L., *Minimization of Waste Solvent: Factors Controlling the Time Between Solvent Changes*, U.S. Army Construction Engineering Research Laboratory (USA-CERL) Contract No. DACA 88-83-C-0012 (Colorado School of Mines, Golden, CO, September 1984).

Castrantas, H. M., R. E. Keay, and D. G. MacKellar, *Treatment of Dry Cleaning Baths*, U. S. Patent 3,677,955 (July, 1972).

Chemical Market Reporter (October 1986).

Development of a Field Test for Monitoring Organic Halides in Waste Fuels, U.S. Environmental Protection Agency (USEPA) Contract No. 68-01-7075 (USEPA, 1968).

Disposal/Recycle Management System Development for Air Force Waste Petroleum Oils and Lubricants, AD 779723 (U.S. Air Force, April 1974).

Donahue, B. A., and M. B. Cramer, *Solvent "Cradle-to-Grave" Management Guidelines for Use at Army Installations*, Technical Report N-168/ADA137063 (USA-CERL, November 1983).

Durrans, T. H., *Solvents*, 2nd ed. (D. Van Nostrand, 1931).

Esposito, G. G., *Solvency Rating of Petroleum Solvents by Reverse Thin-Layer Chromatography*, AD753336 (Aberdeen Proving Ground, 1972).

Evaluation of Chlorine Determination in Waste Oils, USEPA Contract No. 68-01-7075 (USEPA, 1968).

Fisher Scientific Catalog (1986).

Handbook of Chemistry and Physics, 57th ed. (CRC Press, 1976).

International Fabricare Institute (IFT) Bulletin, T-447 (IFT, 1969).

John, A. E., *Drycleaning* (Merrow Publishing Co., Ltd., England, 1971).

Johnson, K., *Drycleaning and Degreasing Chemicals and Processes* (Noyes Data Corporation, 1973).

Kobayashi, H., and B. E. Rittmann, "Microbial Removal of Hazardous Organic Compounds," *Environ. Sci. Technol.*, Vol 16, No. 3 (1982).

Lee, H. J., I. H. Custis, and W. C. Hallow, *A Pollution Abatement Concept, Reclamation of Naval Air Rework Facilities Waste Solvent, Phase I* (Naval Air Development Center, April 1978).

Martin, A. R., and G. P. Fulton, *Drycleaning Technology and Theory* (Textile Book Publishers, 1958).

Mellan, I., *Industrial Solvents Handbook* (Noyes Data Corp., 1977).

Mellan, I., *Industrial Solvents* (Reinhold, 1950).

Memorandum from the Office of the Assistant Secretary of Defense, Director of Environmental Policy, "Used Solvent Elimination (USE) Program," Interim Guidance (February 1985).

Mettler/Paar DMA 35 Density Meter (Mettler Instrument Corp., 1986).

National Institute of Drycleaning Bulletin Service, T-413 (National Institute of Drycleaning [NID], 1965).

Niven, W. W., *Fundamentals of Detergency* (Reinhold, 1950).

Phillips, E. R., *Drycleaning* (NID, 1961).

Randall, C. B., *The Drycleaning Department* (National Association of Dyers and Cleaners, 1937).

Refractometer Manual, ABBE-56 (Bausch and Lomb Optical).

Rosen, J. M. (Ed.), *Structure/Performance Relationships in Surfactants*, ACS Symposium Series 253 (American Chemical Society [ACS], 1984).

Scheflan, L., and M. B. Jacobs, *The Handbook of Solvents* (D. Van Nostrand, 1953).

Shugar, G. J., et al., *Chemical Technician's Ready Reference Handbook*, 2nd ed. (McGraw-Hill, 1981).

Source Assessment: Reclaiming of Waste Solvents, State of the Art, EPA 600/2-78-004f (U.S. Environmental Protection Agency [USEPA], April 1978).

Source Assessment: Solvent Evaporation-Degreasing Operations, EPA-600/2-79-019f (USEPA, August 1979).

Thomas Scientific Catalog (1986-87).

Used Oil and Solvent Recycling Guide (Robert H. Salvesan Associates, 1985).

Willard, H. H., L. L. Merritt, Jr., and J. A. Dean, *Instrumental Methods of Analysis*, 5th ed. (D. Van Nostrand, 1974).

Appendix
Questionnaire Used in Surveying the Field and Responses

Questions

1. Does your facility use Stoddard solvent, PD-680, or mineral spirits? For what? How much? Where is it obtained? Describe the equipment and operating procedures.

2. How do you classify your cleaning operation? Do you require maximal, intermediate, or minimal cleaning?

3. How "clean" do your parts have to be? How is the degree of cleanness determined?

4. How can you tell if the solvent is doing its job?

5. Does (4) above involve any tests? What are the testing procedures? What are the details of the test? Are any written procedures available?

6. How long do you use your solvent? Does the time vary from one operation to another? For the same operation, does the time vary from season to season?

7. How often is the solvent changed? What determines the time for a change? Any makeup added? Are any other chemicals added?

8. Would better tests on when to change solvents be of any use to you? Do you have any problems involving solvent operations? Would increasing the time of solvent use before recycle be helpful?

9. What is done with spent solvent? Is it recycled?

10. What is the nature of the residue (i.e., mainly dirt, oil)?

11. When collecting samples for 2 percent weight test, where in reference to the height of the tank are the samples taken?

12. What is the size of the vats?

13. How deep is the residue at the bottom of the vats?

14. Is the cleaning operation stage-wise or a one-vat job?

15. Is the color of the spent solvent always the same from vat to vat and month to month?

Responses*

Anniston Army Depot

1. Anniston Army Depot buys PD-680 in bulk. The solvent is received in 2400-gal tankers and stored in underground tanks. It is then distributed as needed by a gasoline-type pump system. No information is readily available on where it is obtained. The basic operating procedure is to clean mechanical parts manually in a solvent rinse vat.

2. Parts require minimal cleaning under normal conditions.

3. Parts have to be clean enough to pass a wiping test with a clean white rag.

4. Solvent is ineffective if it leaves a residue or an oil film after drying. An increase in drying time may also be used to decide when to change the solvent.

5. QA inspectors regularly perform a contamination test on samples of solvent by centrifuging. If the solvent contains more than 2 percent solids, it is discarded.

6. The length of time a solvent is used depends on the parts to be cleaned and their condition.

7. Solvent is changed at least once a month and no makeup solvent or performance-improving chemicals are added to spent solvent to extend the usage period.

8. An increase in duration of solvent use is desirable.

9. Spent solvent is used as a fuel along with coal.

10. Residue left on the parts and detected as a result of the wiping test is a combination of dirt and oil.

11. Samples for the monthly test of 2 percent solids are collected by men in the shops. These samples are taken while the parts are being cleaned and are said to be fairly uniform.

12. The vats are approximately 3 ft wide, 4 ft long, and 3 ft deep.

13. There is a residue of varying depth at the bottom of some of the vats. Many of the vats are equipped with filters and do not build up any significant amount of residue.

14. The cleaning operation is a one-vat job, i.e., no stage-wise cleaning is performed.

15. The color variation of spent solvent is significant from shop to shop and month to month.

---

*Numbered responses correspond with numbered questions in the previous section.

325th EMS Wheel and Tire Shop (Tyndall AFB, FL)

1. PD-680 (Stoddard solvent). Used for removal of carbon deposits, grease, and dirt from aircraft wheels and wheel bearings. 700 gal. PD-680 is obtained through normal USAF supply channels. Two each degreaser vats, one used strictly for cleaning aircraft wheel bearings and the other strictly for aircraft wheels. The bearing vat is set in an agitating mode and bearings are cleaned for 30 min. The wheel vat is used for soaking. Soft plastic brushes are used to remove deposits from wheels after the wheels have soaked for at least 5 min.

2. Maximal cleaning.

3. Extremely clean. Through close inspection and experience.

4. The time period it takes to remove hardened deposits from aircraft wheels.

5. No.

6. Varies. Yes. Yes.

7. Depends on how dirty solvent is. Composition of PD-680 is broken down through prolonged use and, consequently, it takes longer to clean aircraft wheels and bearings. No. No.

8. Yes. No. Yes.

9. Disposition through USAF supply channels. Unknown.

10. Nature of residue is mostly dirt and grease.

11. Unknown.

12. 500-gal wheel vat and 200-gal bearing vat.

13. For changing, 1 to 2 in.

14. It is a one-vat job.

15. No. It varies from time to time; depends on how extensively used and length of time involved.

325th EMS Age Branch (Tyndall AFB, FL)

1. PD-680 (Stoddard solvent). Used for removal of grease, dirt, etc., from AGE components prior to disassembly for repair or overhaul. Also used for cleaning internal parts after a component has been dismantled and for equipment wheel bearings prior to repacking. We use an average of 50 gal per month. PD-680 is obtained through normal USAF channels. We use a degreaser vat for all cleaning operations requiring the use of solvent. Cleaning time is contingent on condition of item being cleaned, but normally takes 10 to 30 min per job. Brushes are used to help remove unusually severe deposits.

2. Maximal.

3. Very clean. Through close visual inspection.

4. By the time required for cleaning operations.

5. No.

6. It varies depending on frequency of use. Yes. Yes.

7. It varies depending on how quickly it becomes dirty or diluted. Time for change is based on how dirty it becomes. No. No.

8. Yes. No. Yes.

9. Disposal through USAF supply channels. Unknown.

10. Dirt and grease.

11. Tests not done.

12. 40 gal.

13. Approximately 0.5 to 1 in. when replaced.

14. One vat.

15. No.

Part II

Vapor Degreasing and Precision Cleaning Solvents

The information in Part II is from *Used Solvent Testing and Reclamation, Volume II. Vapor Degreasing and Precision Cleaning Solvents,* prepared by Arthur R. Tarrer, Auburn University; Bernard A. Donahue and Seshasayi Dharmavaram, U.S. Army Construction Engineering Research Laboratory; and Surendra B. Joshi, U.S. Air Force Engineering and Services Center; for the U.S. Army Engineering and Housing Support Center, December 1988.

1. Introduction

Background

Department of Defense (DOD) installations use large amounts of solvent each year in cleaning operations, which generates a huge volume of waste solvent. Much of this waste is or will be considered hazardous as stricter regulations are promulgated and enforced. Thus, proper handling and disposal practices are of increasing concern to DOD. Coupled with these environmental issues is the rising cost of both waste disposal and new solvents. These concerns have prompted DOD to seek safe, cost-effective methods of managing waste solvent.

Solvents used at DOD installations can be classified into five groups based on chemical makeup and function: (1) vapor degreasers, (2) cold-dipping cleaners, (3) paint thinners, (4) paint strippers and carbon removers, and (5) precision cleaners. Most of these solvents are considered to provide one-time use; when they become contaminated, they are discarded. In these cases, disposal methods are mainly destructive, i.e., waste solvents are incinerated, evaporated, or dumped.

Some military facilities have initiated programs for reclaiming used solvents. This option is technically feasible because the solvents usually do not break down chemically during cleaning operations. Their role in cleaning is limited mainly to physical solubilization of waxes, greases, oils, and other contaminants. In fact, laboratory tests of major waste streams at installations have indicated that most solvents present could be recovered by recycling; the reclaimed material would generally be of suitable quality for effective reuse in cleaning.

A situation that has limited recycling is the lack of scientific tests and criteria for judging a solvent as spent (i.e., contaminated to the point that it is no longer effective for its intended purpose). Discarding a solvent before this point fails to maximize the material's life from an economic standpoint, whereas keeping it in service too long may result in use of an ineffective cleaner.

To make solvent recycling practical at installations, DOD needs criteria and simple test(s) for identifying spent solvents and/or indicating when the solvent should be discarded. These tests could have major impact on the environmental and cost issues facing DOD. Specifically, effective tests could:

1. Maximize solvent life for the most economical use of product; solvent cleaning operations would realize a savings through a reduction in new purchases.

2. Allow recycled solvent to be evaluated, preventing the use of inadequate quality materials in the cleaning process.

3. Minimize the amount of hazardous wastes generated, thus limiting the cost of handling and disposal. This benefit is especially important in light of the DOD Used Solvent Elimination (USE) program,[1] which bans the disposal of used solvent in landfills.

Purpose

The overall purpose of this work is to: (1) establish criteria for identifying spent solvents and recommend simple tests to determine when solvents must be changed and (2) evaluate methods of reclaiming solvents as an alternative to disposal.

Volume I reported on the first phase of this work and addressed Stoddard-type solvents used in cleaning operations. Volume II covers the second phase--halogenated compounds used in vapor degreasing and metal cleaning/surface preparation. In addition to the overall purpose, specific objectives of this phase were to: (1) analyze chlorinated solvent inhibitors and determine relationships between inhibitors and usage time, (2) observe reactions between inhibitors and degreaser acids, and (3) evaluate halogenated solvent reclamation methods and determine if inhibitors are lost in chlorinated solvent reclamation.

Approach

This phase of the study involved the following steps:

1. Review the literature for methods of monitoring solvent quality.

2. Investigate physicochemical test methods.

3. Analyze halogenated solvents and evaluate test results.

4. Rate the test methods and determine their practicality for use in the field.

5. Evaluate methods of reclaiming spent and partially spent solvent.

6. Study the economics of solvent reclamation.

7. Recommend test methods and reclamation strategies for use at military installations.

Scope

This report addresses vapor degreasing and precision cleaning solvents. The vapor degreasing solvents investigated in this study were trichloroethylene (TCE), tetrachloroethylene (perchloroethylene, PERC), and 1,1,1-trichloroethane (methyl chloroform, MC). Metal preparation and precision cleaning solvents studied were isopropanol (IPA) and 1,1,2-trichloro-1,2,2-trifluoroethane (freon-113).

---

[1]Memorandum from the Office of the Assistant Secretary of Defense, Director of Environmental Policy, "Used Solvent Elimination (USE) Program," Interim Guidance (February 1985).

Mode of Technology Transfer

It is recommended that the test procedures be verified in the field and refined through a transfer medium such as the Facilities Engineering Applications Program (FEAP). When the tests are validated, they should be implemented at all military installations where solvents are used. The procedures will be incorporated into the appropriate technical manuals for implementation.

2. Solvent Use and Characterization

Halogenated solvents are used mainly in vapor degreasing and metal preparation/precision cleaning. For background information, this chapter describes:

- Mechanisms of the cleaning process.

- Solvent characterization.

- Use and generation of halogenated solvent waste at military facilities.

Cleaning Mechanism

To understand how solvents work, it is first necessary to look at the properties of contaminants that require solvent cleaning. Solvent contaminants are generally heterogeneous mixtures of substances with different physical and chemical characteristics. They can be grouped roughly as: (1) hydrocarbon oils, such as lubricating oils, greases, and tar; (2) paints and varnishes; and (3) soily material such as clay, silt, cement, soot, and lampblack.[2] The first two categories are inert organic materials (mostly liquids), whereas the third category consists mainly of insoluble inorganic materials that are solids in various states of subdivision.

Vapor degreasers are used primarily for removing oils and greases that are soluble in the degreasing solvent. The adhesion of the oily, greasy contaminant to a metal or plastic part (work) is through bonding to the work by cohesion or wetting. Work to be cleaned is immersed in the vapor zone of a degreaser. Since the work is introduced at ambient temperature, the solvent vapor condenses on the cooler exposed surface of the part and dissolves the contaminants. This cleaning action continues until the work reaches the vapor temperature. The degree of solubility of a solute (contaminant) in a solvent is known as the "solvent power" of the solvent. The amount of solvent vapor condensation on a part's surface depends on its weight and specific heat.

Solvent Characterization

Several characteristics are required of cleaning solvents.[3] They must:

- Dissolve oils, greases, and other contaminants.

- Have a high vapor density relative to air and a low vapor pressure to minimize solvent losses.

- Be chemically stable under conditions of use.

- Be noncorrosive to common materials of construction.

---

[2] W. W. Niven, *Fundamentals of Detergency* (Reinhold, 1950).

[3] U.S. Environmental Protection Agency (USEPA), *Source Assessment: Solvent Evaporation-Degreasing Operations*, EPA-600/2-79-019f (Industrial Environmental Research Laboratory, August 1979); W. W. Niven.

● Have a low boiling point and latent heat of vaporization so that the solvent can be separated from oil, grease, and other contaminants by simple distillation.

● Not form azeotropes with liquid contaminants or with other solvents.

● Be available at reasonable cost.

● Remain nonexplosive and nonflammable under the operating conditions.

In a cleaning operation, the choice of solvent depends on the requirements of the cleaning process (i.e., solvent compatibility with part and soil, necessary boiling and vapor characteristics, cost of operation, solvent stability, toxicity requirements, and method of handling parts).[4]

Solvents used at DOD installations can be classified as (1) vapor degreasers, (2) cold-dipping cleaners, (3) paint thinners, (4) paint strippers and carbon removers, and (5) precision cleaners. Vapor degreasing solvents are mostly chlorinated compounds (e.g., 1,1,1-trichloroethane, trichloroethylene, and tetrachloroethylene). Mineral spirits and Stoddard solvent are used as cold-dipping solvents (see Volume I). Paint thinners are generally oxygenated compounds (e.g., methyl ethyl ketone) and alcohols, along with toluene and xylene. Paint strippers and carbon removers contain methylene chloride blended with additives. Metal preparation and precision cleaning solvents include alcohols and freons.[5] The solvents addressed in this report are characterized briefly below.

Trichloroethylene (TCE)

TCE (C_2HCl_3) is the chlorinated solvent traditionally used in industrial cleaning. It is a stable, colorless liquid and can be vaporized with low-pressure steam (135.7 to 204.6 kPa) because of its low boiling point (87.2 °C). It has an aggressive solvent action and manufacturing costs are low. Its efficient action leaves no residue to interfere with subsequent metal treatment or finishing.[6] At present, however, TCE use is limited due to Occupational Safety and Health Administration (OSHA) standards for worker exposure (see Table 1).

Tetrachloroethylene (Perchloroethylene, PERC)

PERC (C_2Cl_4) has been used for many years in specific cleaning operations. It is a colorless liquid and has a boiling point of 121.1 °C. It is highly resistant to breakdown under heavy workloads and adverse working conditions. PERC is also used in removing

---

[4] USEPA, August 1979; W. W. Niven.

[5] USEPA, August 1979; USEPA, *Source Assessment: Reclaiming of Waste Solvents, State of the Art*, EPA 600/2-78-004f (Industrial Environmental Research Laboratory, April 1978); A. L. Bunge, *Minimization of Waste Solvent: Factors Controlling the Time Between Solvent Changes*, CERL Contract DACA 88-83-C-0012 (Colorado School of Mines, September 1984); R. W. Bee and K. E. Kawaoka, *Evaluation of Disposal Concepts for Used Solvents at DOD Bases*, TOR-0083(3786)-01 (The Aerospace Corp., February 1983); H. J. Lee, I. H. Custis, and W. C. Hallow, *A Pollution Abatement Concept, Reclamation of Naval Air Rework Facilities Waste Solvent, Phase I* (Naval Air Development Center, April 1978).

[6] USEPA, August 1979; W. W. Niven; T. J. Kearney and C. E. Kircher, "How To Get the Most From Solvent-Vapor Degreasing--Part I," *Metal Progress* (April 1960), pp 87-92.

high-melting-point waxes and in spot-free drying of metal parts because of its high boiling point.[7] However, the high temperature can damage certain materials, including plastics. It is the third largest volume vapor degreasing solvent, with 43,000 metric tons (MT) used every year.[8] PERC use, like that of TCE, is limited by OSHA standards (see Table 1).

1,1,1-Trichloroethane (Methyl Chloroform, MC)

MC (CH_3CCl_3) has recently found increasing use in metal cleaning operations due to the OSHA limitations placed on TCE and PERC. It is the largest volume vapor degreasing solvent with more than 90,000 MT being consumed annually. It has a lower toxicity than TCE or PERC (Table 1) and is an excellent solvent for many oils, greases, and waxes. MC is an excellent solvent for cleaning plastics, polymers, and resins because it is much less likely to degrade these substances than are TCE and PERC. Its use, however, is limited because of its tendency to hydrolyze and to form acid byproducts when boiled in the presence of water. MC reacts violently with aluminum and some other metals (e.g., zinc, magnesium) and must be stabilized to prevent such actions. Also, exposure of MC to strong alkalies must be avoided.[9]

Table 1

Exposure Limits of Chlorinated Solvents*

Solvent	OSHA Exposure Limits (ppm)		ACGIH-TLV** (ppm)		
	8-Hr Time-Weighted Avg.	Acceptable Ceiling Conc.	Acceptable Max. Peak Weighted Avg. (TWA)	TLV	Short Time Exposure Limit (STEL)
PERC	100	200	300, 5 min in any 3 hr	50	200
TCE	100	200	300, 5 min in any 2 hr	50	200
MC	350	No Limit	No Limit	350	450

*Source: *Degreasing* (Dow Chemical Company, 1985). Used with permission.
**American Conference of Governmental Industrial Hygienists (ACGIH) recommended Threshold Limit Value (TLV).

---

[7] USEPA, August 1979; T. J. Kearney and C. E. Kircher, April 1960; T. J. Kearney and C. E. Kircher, "How To Get the Most From Solvent-Vapor Degreasing--Part II," *Metal Progress* (May 1960), p 93.

[8] USEPA, August 1979.

[9] USEPA, August 1979; R. Monahan, "Vapor Degreasing With Chlorinated Solvents," *Metal Finishing* (November 1977), pp 26-31; R. L. Marinello, "Metal Cleaning Solvents," *Plant Engineering* (30 October 1980), pp 50-57; *ASTM Handbook of Vapor Degreasing*, ASTM Special Publication No. 310 (American Society for Testing and Materials [ASTM], April 1962).

Isopropanol (Isopropyl Alcohol, IPA)

IPA is a colorless liquid with a boiling point of 82.2 °C. It is used as a metal and plastic preparation solvent and also sometimes as an extender in paint strippers. Precision cleaning operations include the maintenance of fragile, sensitive navigational and electronic equipment.

1,1,2-Trichloro-1,2,2-trifluoroethane (Freon-113)

Freon-113 is used to clean specific electrical and navigational instruments and parts that require a solvent with high solvency power and rapid evaporation rate. Freons generally have high density (1.5 times that of water), low viscosity, low surface tension, and low boiling point.[10]

Generation of Waste Solvent

Roughly 25,000 vapor degreasing operations in the United States use halogenated solvents, including freons. The annual consumption of TCE, PERC, and MC in vapor degreasing operations amounts to 112,700, 43,000, and 90,000 MT, respectively.[11] The breakdown of nationwide freon and IPA consumption as metal preparation and precision cleaning solvent is not available. However, the overall annual freon (including all freons) and IPA consumption in the United States exceeds 428,600 and 803,000 MT, respectively.[12]

Military installations within DOD can be classified as large or small. Large bases include shipyards, air logistic centers, Army depots, and air rework facilities. This category of DOD installation generates 400 drums (55 gal each) of spent chlorinated solvents annually from vapor degreasing operations, and there are 29 such installations. Small installations are much more numerous (approximately 124 bases are considered in the study by Bee et al.) and are low-volume solvent consumers. IPA requirements as metal preparation/precision cleaning solvent are in small quantities and are generally consumed in process. Waste freon generation is about 75 drums annually at an installation using it. Table 2 lists amounts of spent vapor degreasing and metal preparation/precision cleaning solvents generated at some DOD bases.[13]

A pollution abatement study conducted by Lee et al. reveals that a substantial volume of solvents is being used annually by five of the six Naval Air Rework Facilities (NARFs).[14] Table 3 lists the amount used by each NARF.

The various disposal alternatives for all waste solvent generated at DOD installations were reviewed in Volume I.

---

[10]R. Monahan; R. L. Marinello.
[11]USEPA, August 1979.
[12]USEPA, August 1979; *ASTM Handbook of Vapor Degreasing.*

[13]R. W. Bee and K. E. Kawaoka.
[14]H. J. Lee, I. H. Custis, and W. C. Hallow.

Table 2

Spent Solvent Generation at Major DOD Bases*

Bases	Vapor Degreasing Solvents	Metal Prep/ Precision Cleaning Solvents	All Solvents: Total
Seneca Army Depot, NY	15**	None	180
Kelly AFB, TX***	700	62	1134
Hill AFB, UT	545	10	2270
Tyndall AFB, FL	None	None	118
Jacksonville NAS, FL	460	113	2285
Davis Monthan AFB, AZ	3	None	227
Bergstrom AFB, TX	None	None	243
Corpus Christi Army Depot, TX	275	NA	1025
Norfolk NARF, VA	100	NA	1084
McClellan AFB, CA	150	75	935
Robins AFB, GA	700	70	870

*Source: R. W. Bee and K. E. Kawaoka, *Evaluation of Disposal Concepts for Used Solvent at DOD Bases*, Contract No. F04701-82-C 0083 (The Aerospace Corporation, February 1983). Used with permission.
**All quantities are reported as 55-gal drums/year.
***AFB = Air Force Base; NAS = Naval Air Station; NA = not applicable.

Table 3

Spent Solvent Generated at Naval Air Rework
Facilities (NARFs) (in 1000 gal/hr)*

Base	TCE	PERC	MC	IPA	Freon-113	All Solvents: Total
Alameda, CA	-	-	24.2	1.2	9.9	228.2
Norfolk, VA	-	-	5.6	0.1	26.3	98.8
North Island, CA	-	-	48.0	-	9.7	413.9
Pensacola, FL	7.3	-	31.0	-	10.6	243.5
Jacksonville, FL	60.0	0.1	7.4	-	-	118.7
Total	67.3	0.1	116.2	1.3	56.5	1103.1

*Source: H. J. Lee, I. H. Custis, and W.C. Hallow, *A Pollution Abatement Concept, Reclamation of Naval Air Rework Facilities Waste Solvent, Phase I* (Aircraft and Crew Systems Technology Directorate, Naval Air Development Center, April 1978).

3. Development of Detection Methods

Literature Review

Volume I contains a comprehensive literature survey of physicochemical methods for monitoring solvent quality. All test methods that were selected to monitor Stoddard solvent also apply to vapor degreasing and precision cleaning solvents. In this chapter, only those tests specific to halogenated solvents are reviewed.

The American Society for Testing and Materials (ASTM) has several standard tests on chlorinated organic solvents.[15] A brief outline of those relevant to halogenated compounds is given below.

ASTM Standard D 2106-78 on amine acid acceptance method measures the concentration of an amine (basic) inhibitor by titration with standard acid.

ASTM Standard D 2942-74 on total acid acceptance value (AAV) method measures the total concentration of amine and neutral-type (alpha epoxide) inhibitors in a solvent.

ASTM Standard D 2989-74 on acidity-alkalinity determines the acidity in halogenated solvents. This test is done either by using a glass electrode pH meter or an indicator (bromothymol blue) and titration with a standard solution of sodium hydroxide in anhydrous methanol. The alkalinity is determined using ASTM D 2106.

ASTM Standard D 2943-76, aluminum scratch test for MC, determines if adequate solvent degradation inhibitor is present in MC. A cleaned coupon of aluminum is immersed in the solvent and scratched. After enough time is allowed for a reaction, the presence or absence of bubbling, solvent discoloration, or dark residue is correlated with inhibitor strength.

ASTM Standard D 2108-71 on color measurement checks for visible turbidity in "water-white" halogenated solvents using a color comparator.

ASTM Standard D 2251-67 on metal corrosion evaluates halogenated solvents corrosivity to metals. This test serves as a guide for selecting an appropriate solvent to clean metal parts. Polished metal strips are immersed in a solvent and heated at reflux temperature for 60 min. The strips are then inspected for evidence of corrosion.

ASTM Standard D 2109-78 on nonvolatile matter determines the nonvolatile matter content by evaporating a known amount of solvent in an oven at 105 °C.

ASTM Standard D 2110-78, test method for pH of water extractions of halogenated solvents, requires contacting the solvent with freshly boiled, distilled water, followed by the measurement of pH using either a glass electrode pH meter or a Gramercy pH indicator.

ASTM Standard D 2250-67 on physical and electrical breakdown of insulating materials by halogenated organic solvents determines the physical and electrical changes that occur in insulating materials during immersion in certain solvents at room temperature.

---

[15] *Annual Book of ASTM Standards* (1980).

This test is used to select solvents for cleaning electrical insulating materials and equipment.

ASTM Standard D 1901-67 on relative evaporation time of solvents measures the time required for complete evaporation of a thin film of solvent from a sheet-metal panel in comparison with a reference standard solvent. The test gives valid comparisons of the evaporation rates of different solvents in order to determine the best solvent for a cleaning application.

ASTM Standard D 2111-71 determines specific gravity of a solvent by using either a hydrometer, specific gravity balance, or pycnometer.

ASTM Standard D 3401-78 determines water content electrometrically within the concentration range of 5 to 500 ppm in halogenated solvents using Karl-Fischer reagent.

Arbitrary criteria to identify a spent solvent have evolved in various industries; for example, in the drycleaning industry, the transmittance of light through a sample of dirty solvent determines when to change the solvent.[16] In other degreasing operations, solvent color as well as the presence of dirt and grease are taken into consideration for changing solvents.[17]

Another category of tests (as compared with those based on physical parameters) rates solvent power in terms of chemical reactivity or solubility of certain materials. These tests include acid number, aniline point, dimethyl sulfate value, Kauri-butanol value (KBV), and cellulose-nitrate solution value.[18] The KBV has traditionally been used in the drycleaning and varnish industries to represent solvent performance. The KBV of a solvent decreases with increased contamination of the solvent due to grease, oil, or soil.

Physicochemical Tests

The experimental study on chlorinated solvents consisted of measuring the following physicochemical properties: (1) KBV, (2) viscosity, (3) specific gravity, (4) refractive index, (5) visible light absorbence, (6) electrical conductivity, (7) AAV, and (8) boiling point. These techniques were selected from the literature on the basis of reported scientific reliability and consistency and because they would be relatively easy to perform in the field. Also, in phase I of this project, tests 1 through 6 yielded consistent results (see Volume I of this report). The background and procedures for these tests are described below.

---

[16]K. Johnson, *Drycleaning and Degreasing Chemicals and Processes* (Noyes Data Corp., 1973); *International Fabricare Institute (IFI) Bulletin*, T-447 (1969); *National Institute of Drycleaning (NID) Bulletin Service*, T-413 (1965); H. M. Castrantas, R. E. Keay, and D. G. MacKellar, U.S. Patent 3,677,955 (July 1972).
[17]A. L. Bunge.
[18]*Annual Book of ASTM Standards*; IFI Bulletin 447.

Kauri-Butanol Value (KBV)

The KBV method evaluates the relative power of solvents. This method gives an index for ranking solvents on their ability to dissolve other materials.[19] The paraffinic hydrocarbons have the lowest solvent power whereas the polar and aromatics have the highest. Therefore, the solvent power in terms of KBV is largely dependent on the amount of aromatics and the polarity of the solvent. This simple test is widely accepted as a good measure of relative solvent power. The apparatus and reagents required and the ASTM test procedure are described below.

Apparatus.

● Erlenmeyer flask, 250-mL.

● Buret, 50-mL (Figure 1).

● Print sample--a sheet of white paper having on it black 10-point print, No. 31 Bruce Old Style type.

Reagents.

● Standard Kauri-butanol (K-B) solution; a prepared solution was obtained from Chemical Service Laboratory, 5543 Dyer St., Dallas, TX, 75206.

● Reagent-grade toluene.

● Reagent-grade heptane.

Figure 1. Kauri-butanol value apparatus.

---

[19]*Annual Book of ASTM Standards*; E. R. Phillips, *Drycleaning* (NID 1961); G. G. Esposito, *Solvency Rating of Petroleum Solvents by Reverse Thin-Layer Chromatography*, AD-753336 (Aberdeen Proving Ground, 1972).

Procedure.[20]

1. Weigh 20 g of standard K-B solution into a 250-mL Erlenmeyer flask.

2. Place the sheet containing 10-point print under the flask.

3. Fill buret with solvent to be sampled.

4. Titrate into the flask until printed material becomes obscured or blurred but not to the point where the print becomes illegible.

5. Calculate the KBV using Equation 1:

$$KBV = 65(C-B)/(A-B) + 40 \qquad\qquad [Eq\ 1]$$

where   A = amount of toluene (mL) required to titrate 20 g of K-B solution (should be around 105);   B = amount of 75 percent heptane/25 percent toluene blend needed to titrate 20 g of K-B solution (should be around 40 mL); and C = amount of sample solvent (mL) needed to titrate 20 g of K-B solution. Values of A and B can be obtained from standard titrations or from the K-B solution manufacturer.

Viscosity

Viscosity is the internal friction or resistance to flow that exists within a fluid, either liquid or gas. This property depends on the intermolecular attractive forces within the fluid.

Viscosity is an extremely useful method for characterizing oils and solvents. Viscosities of "heavier" and "lighter" oils are significantly different, whereas their densities may differ very little.

A common unit of viscosity is the poise, which is equal to 1 gram per centimeter second (g/cm-sec) and is usually tabulated in centipoise (cp). Viscosities in this work were measured using an Ostwald viscometer. This type of viscometer measures the flow rate of a fluid through a capillary tube in a gravity field. Newtonian behavior was assumed for all solvent mixtures. A Newtonian fluid is one that shows a linear relationship between the magnitude of an applied shear stress and the resulting rate of deformation. The viscosity (μ) of a given fluid is calculated using Equation 2:

$$\mu = k \cdot t \qquad\qquad [Eq\ 2]$$

where t is the time required for a fixed volume of fluid to flow through the capillary and k is a constant obtained by measuring the time of a liquid having a similar known viscosity.[21]

---

[20]E. R. Phillips.
[21]G. J. Shugar, et al., *Chemical Technician's Ready Reference Handbook*, 2nd ed. (McGraw-Hill, 1981).

The apparatus and procedure for this test are as follow.

Apparatus:

- Ostwald viscometer (Figure 2).

- Stopwatch.

Procedure.[22]

1. Wash the viscometer thoroughly and rinse with distilled water, making sure the instrument is clean and dry before taking readings.

2. Introduce distilled water and allow it to come to thermal equilibrium in a constant-temperature bath.

3. Using a suction bulb, draw liquid into the upper bulb to the marked line.

4. Remove the bulb and record the time needed for the level of water to pass between markings.

5. Repeat steps 3 and 4 until readings are fairly constant.

6. Clean and dry viscometer thoroughly.

7. Add an appropriate volume of the solvent to be tested to the viscometer.

8. Repeat steps 3 through 5.

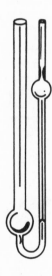

Figure 2. Ostwald viscometer.

[22]G. J. Shugar, et al.

Refractive Index

The refractive index of a liquid is the ratio of the velocity of light in a vacuum to the velocity of light in the liquid. This property can be used to identify a substance and determine its purity. Since the angle of refraction varies with the wavelength of light used, the measurement of refractive index requires that light of a known wavelength be used. However, a white light can be used if the refractive index of a reference liquid is measured in the same light.

Refractive index is commonly reported to four decimal places, and since it can be determined experimentally to a few parts in 10,000 easily, it is a very accurate physical constant. Small amounts of impurities can have significant effects on the experimental value.

Refractive indices in this study were determined using an Abbe refractometer. This device compares the angles at which light from an effective point source passes through a test liquid and into a prism whose refractive index is known.

The procedure for measuring the refractive index of a liquid is as follows.[23]

1. Unlock the hinged assembly and lower the bottom part of the prism.

2. Clean the upper and lower prisms with soft, nonabrasive, absorbent, lint-free cotton wetted with benzene. Rinse by wiping with petroleum ether and allow to dry.

3. Place a drop of solution of known refractive index (water) on the prism.

4. Record the temperature indicated by the thermometer next to the prism.

5. Set the scale to correspond with the known refractive index at the corresponding temperature.

6. Look through the eyepiece and turn the compensator knob until the colored indistinct boundary seen between the light and dark fields becomes a sharp line.

7. Adjust the magnifier arm until the sharp line exactly intersects the midpoint of the crosshairs in the image (Figure 3).

8. Repeat steps 1 through 7 using the solvent to be tested.

9. Clean the prisms and lock them together.

Specific Gravity

Specific gravity is defined as the ratio of the density of a liquid to that of water at the same temperature. Density is a fundamental physical property of a substance denoting the mass of a substance per unit volume.

Specific gravity can be measured easily using a specific gravity meter, hydrometer, or pycnometer. Any of these devices can give very accurate results with little training or experience. However, a hydrometer is generally not sensitive enough to

---

[23] *Abbe-56 Refractometer Manual* (Bausch and Lomb Optical Co.).

detect the small specific gravity variations that occur when a solvent becomes contaminated.

In this study, specific gravity was measured by two methods. The first method used a pycnometer bottle which holds a precise volume of liquid and is weighed on a balance. The second method used an electronic specific gravity meter (Mettler/Paar DMA35SG).[24]   Both methods gave accurate results, but the specific gravity meter was easier to use and requires no weighing. Specific gravity is very sensitive to changes in temperature because, as the temperature increases, a fluid will have a tendency to expand, thus reducing the amount of mass in the same volume of fluid. The opposite effect occurs when a fluid is cooled.

The apparatus and procedures for the two test methods are as follow.

<u>Apparatus</u>.

● Analytical precision balance.

● Pycnometer (Figure 4), or

● Electronic specific gravity meter.

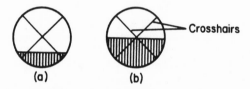

(a)   (b)

Figure 3.   Adjustment of the refractometer: (a) incorrect (b) correct.

Figure 4. Pycnometer.

---

[24] *Mettlar/Paar DMA 35 Density Meter* (Mettler Instrument Corp., 1986).

Procedure. For the pycnometer:[25]

1. Clean pycnometer thoroughly.

2. Dry pycnometer in an oven for 30 min.

3. Remove from oven and allow to cool.

4. Weigh pycnometer on a precision balance.

5. Repeat steps 2 through 4 until a constant weight is obtained.

6. Fill pycnometer with liquid completely.

7. Wipe cap with tissue and weigh pycnometer.

When using a specific gravity meter:[26]

1. Turn on meter.

2. Fill bulb on meter, making sure no air bubbles are in the measuring tube as this will cause errors.

3. Record the temperature of the liquid as well as the specific gravity referenced to 20 °C.

4. Turn off meter.

5. Empty meter of all fluid and clean thoroughly.

Electrical Conductivity

According to Ohm's Law, the resistance of a conductor of length L and cross sectional area A is given by:

$$R = k \times (L/A) \qquad [Eq\ 3]$$

where R is the resistance in ohms and k is the specific resistivity, a property of the material being examined (expressed in ohm-cm). In dealing with liquids, the usual practice is to measure the reciprocal of k, called the specific conductance or conductivity, expressed in ohm-1cm-1 or mhos/cm. Thus, from Equation 3, the conductivity is given by:

$$c = (L/A) \times (1/R) \qquad [Eq\ 4]$$

Electrical conductivity is generally measured using a conductance cell for which the factor L/A appearing in Equation 4 can be determined by measuring the (known) conductance of a standard solution, usually potassium chloride in water. For a cell of given geometry, the factor L/A is called the "cell constant" and, once it has been determined, the conductance of unknown solutions can be extrapolated by applying the same procedures.

---

[25]G. J. Shugar, et al.
[26]*Mettler/Parr DMA 35 Density Meter.*

The electrical conductivity of pure organic liquids is usually very small--on the order of 10^{-8} mhos/cm or less at 25 °C. Electrical conductivity is a function of temperature so that some attention must be given to controlling a liquid's temperature during a conductivity determination.

Since the conductance of a solution is a function of concentration, it would be expected that the electrical conductivity of a solvent would change as impurities are accumulated during usage. Thus, monitoring a solvent's conductivity might provide one method of indicating solvent bath contamination. During the course of solvent usage, its conductivity could increase or decrease, depending on the particular impurities being accumulated.

The conductivity test apparatus and procedure are described below.

Apparatus.

● Conductivity meter (YSI Model 32).

● Probe (YSI #3402).

Procedure.

1. Clean probe thoroughly (Figure 5).

2. Measure temperature of the solvent to be tested.

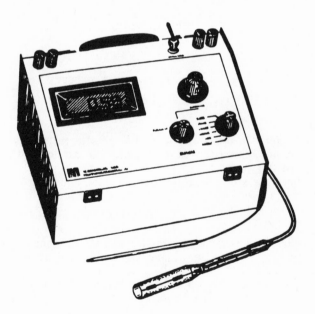

Figure 5. Conductivity meter with probe.

3. Set conductivity meter to conductivity setting.

4. Dip probe in solvent and set instrument to proper scale.

5. Allow about 5 min for probe to come to equilibrium.

6. Record conductivity.

7. Repeat until constant.

Visible Absorbence Spectrometry

There is usually a noticeable change in the color of a contaminated solvent; therefore, visible absorbence is an obvious choice for a test method. The amount of light absorbed could indicate the solvent's level of contamination. This type of test is currently used in the drycleaning industry.[27]

When an electromagnetic wave of a specific wavelength impinges upon a substance, the fraction of the radiation absorbed will be a function of the concentration of substance in the light path and the thickness of the sample. It has been found that increasing the concentration of the absorber has the same effect as a proportional increase in the radiation-absorbing path length (Beer's law). Therefore, the absorbence A is proportional to the concentration of absorbing solute:[28]

$$A = a \cdot b \cdot c \qquad\qquad [Eq\ 5]$$

where a is specific absorptivity in $g^{-1}.cm^{-1}$, b is the sample path length in cm, and c is the solvent concentration in g/L. This equation holds only for low concentrations. The derivation of Beer's law assumes the use of monochromatic light. However, if absorptivity is essentially constant over the instrumental bandwidth, Beer's law will be followed closely. Departure from Beer's law is most serious for wide slit widths and narrow absorption bands and is less significant for broad bands and narrow slits. Therefore, the most significant measurements in this study were made using a very narrow slit width of 0.02 mm and a broad band between 400 and 600 nm.

The apparatus and test procedure for measuring visible absorbence are as follow:

Apparatus.

• UV/visible spectrophotometer (Gilford 250).

• Cuvette.

Procedure.

1. Turn on instrument and allow a 30-min warmup.

---

[27]K. Johnson; I. Mellan, *Industrial Solvents* (Reinhold, 1950).
[28]H. H. Bauer, G. D. Christian, and J. E. O'Reilly, *Instrumental Analysis* (Allyn and Bacon, 1979); H. H. Willard, L. L. Merritt, Jr., and J. A. Dean, *Instrumental Methods of Analysis*, 5th ed. (D. Van Nostrand, 1974).

2. Clean and dry cuvette, making sure all smudges are wiped off and that there are no scratches on any surfaces in the light path.

3. Fill cuvette with sample and place in the spectrometer.

4. Set slit width and take reading.

Thin-Layer Chromatography

Chromatography, in general, is a separation technique based on the fact that a substance has different affinities for each of two phases--stationary and mobile. The relative distribution of a substance between the two phases is known as the distribution coefficient K. The fact that different substances have different distribution coefficients leads to the possibility of separation by chromatography. Two substances, A and B, with unequal distribution coefficients K_A and K_B, will spend different amounts of time in the mobile and stationary phases. Movement of the mobile phase leads to a separation. For example, if $K_A > K_B$, then A spends more time in the mobile phase and thus travels faster than substance B.[29]

In classic thin-layer chromatography (TLC), a mixture to be separated (in this case, a mixture of dyes) would be deposited at a starting point on a plate or paper and the mobile phase (solvent) would be allowed to travel up the plate/paper by capillary action. Because of differences in distribution coefficients, the substances (dyes) in the mixture separate. The distances traveled by substances A and B, and the solvent (mobile phase), respectively, would then be recorded. The separation efficiency is presented as R_f (response factor) values, where R_f is the ratio of the distance traveled by a substance being separated to that of the solvent (mobile phase).

The ability of a solvent to keep a substance in solution can be termed its "solvent power" with respect to that substance. Visualizing a solvent as the mobile phase in TLC, it is clear that a solvent with higher solvent power should give rise to a higher K_A and higher R_f value than with lower solvent power. TLC has been used by Esposito to rank solvent power of various petroleum solvents.

The supplies and procedure are as follow.

Apparatus.

- TLC glass microfiber paper.

- Dyes (Brilliant Blue and Disperse Yellow 9).

Procedure.

1. Activate paper by heating in an oven for 30 min.

2. Cut paper into strips that will fit into the TLC jar.

3. Place a mark across the paper approximately 2 cm from the bottom.

---

[29]G. G. Esposito; H. H. Bauer, G. D. Christian, and J. E. O'Reilly; H. H. Willard, L. L. Merritt, Jr., and J. A. Dean.

4. Pour solvent to be tested into the jar to a level of about 1 cm.

5. Spot dye(s) on paper at the 2-cm mark.

6. Place paper into jar so that it remains vertical. The dye spots should be about 1 cm above liquid level.

7. Cover the jar so that the liquid and vapor can come to equilibrium.

8. When the solvent (mobile) phase nears the top of the strip, remove strip from bottle and mark solvent front (Figure 6).

9. Allow strip to dry.

10. Measure the distance traveled by the dyes from the starting line to the middle of the dye spot. Then measure the distance traveled by the solvent front. The ratio of the distance traveled by the dye to that of the solvent is the R_f value.

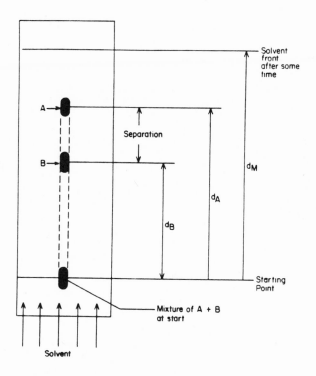

Figure 6. Mechanism of thin-layer chromatography.

Acid Acceptance Value (AAV)

The AAV (ASTM Standard D 2942)[30] is the most common method presently used to determine the inhibitor level of chlorinated solvents. It measures the acid acceptance inhibitor level present in these solvents. The total acid acceptance is determined by reaction with nonaqueous hydrochloric acid (HCl) in excess. The excess acid is then neutralized with a standard sodium hydroxide (NaOH) solution.

The procedure is as follows:

1. Pipet 25 mL of hydrochlorinating agent (0.1 N HCl in isopropyl alcohol) and 10 mL of isopropyl alcohol into a 250-mL Erlenmeyer flask.

2. Add 3 drops of bromophenol blue indicator solution (1 g bromophenol blue in 800 mL water and 200 mL denatured ethanol) and titrate to a stable endpoint with 0.1 N NaOH solution.

3. Pipet 25 mL of hydrochlorinating solution into a glass-stoppered Erlenmeyer flask. Add 10 mL of chlorinated (halogenated) solvent and 25 mL of isopropanol.

4. Shake thoroughly and allow to stand at room temperature for 10 min. Add 3 drops of bromophenol blue indicator solution to the flask and titrate to a stable endpoint with 0.1 N NaOH solution.

5. Calculate the acid acceptance as weight percent NaOH:

$$AAV \text{ (wt\% NaOH)} = [(A-B)N \cdot 0.04 \cdot 100]/W \qquad \text{[Eq 6]}$$

where: A = mL NaOH solution required for titration of blank.

B = mL NaOH solution required for titration of sample.

N = normality of the NaOH solution.

W = amount of solvent sample used (g).

In Equation 6, the constants 0.04 and 100 are the conversion factors from mL equivalent/L NaOH to g NaOH, and weight fraction Na to weight percent NaOH, respectively.

Sampling

Samples of vapor degreasing solvents were obtained from DOD installations that typically do metal cleaning and from Hayes International, Birmingham, AL, a DOD contractor in private sector (Table 4). Samples included new solvent, samples taken at set time intervals within a vapor degreasing operation cycle (from startup to cleanout), and spent solvent.

The MC gave erratic results because of prolonged use of solvent and several makeups. These MC samples were not evaluated further except to study the effect of

---

[30] *Annual Book of ASTM Standards.*

reclamation (see Chapter 4). Therefore, the **Test Results** section below is limited to solvents and samples obtained from the DOD installations only.

Test Results

New and spent chlorinated solvents were analyzed by the test methods described earlier in this chapter. The results are presented below.

TCE

TCE was obtained from a degreaser at various time intervals beginning on the day a new solvent was placed into the degreaser and ending on the day the solvent was changed. Six such samples were provided. The solvent residence time in the degreaser was 12 days.

The tests conducted on TCE were KBV, viscosity, refractive index, specific gravity, electrical conductivity, visible absorption spectrometry, solids content, AAV, and TLC. New makeup solvent was introduced whenever the solvent level in a degreaser became low. This addition is reflected as sudden increase or decrease in the experimental data. A brief discussion of the test results follows.

KBV. Figure 7 shows the variation in KBV of TCE with usage period. As in the case of Stoddard solvent, the KBV did not decrease significantly with use (<2 percent). Because of the small change in the KBV over the usage life of the solvent, this test is not very feasible as an effective test to monitor and predict solvent change. However, for reclaimed solvents, this test can be used to monitor the solvent power.

Viscosity. The variation in viscosity of TCE with usage time is shown in Figure 8. The increase in viscosity was significant (10 percent) over the entire range of days used. However, the increase in viscosity was not as marked as in the case of Stoddard solvent.

Table 4

Collection Sites for Solvents Used in This Study

Cleaning Operation	Solvent
Robins AFB, Warner-Robins, GA	Trichloroethylene 1,1,1-Trichloroethane Isopropanol Trichlorotrifluoroethane
Kelly AFB, San Antonio, TX	Tetrachloroethylene
Hayes International Corp., Birmingham, AL	1,1,1-Trichloroethane

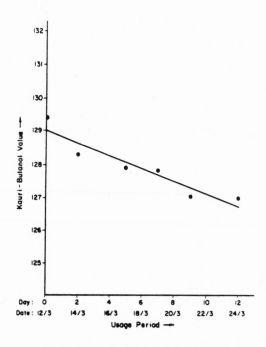

Figure 7. Variation in Kauri-butanol value of trichloroethylene with usage time.

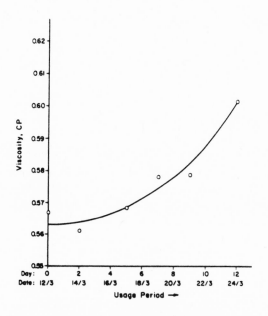

Figure 8. Variation in viscosity of trichloroethylene with usage time.

The procedure for this test is fairly simple and can be performed in the field with minimal training of personnel.   However, some form of temperature control (e.g., a water bath) is required to ensure that the experimental data are interpreted correctly.   A curve representing the temperature effect could be constructed for measurements at different temperatures as an alternative to temperature control.   Also, filtration of the solvent may be necessary if the particulate content is high.

Refractive Index.   This value was measured using a Bausch and Lomb Model Abbe-56 refractometer.[31]   The refractive index of water was measured to be 1.3309 at 30 °C as compared with a literature value of 1.3319.[32]   The refractive index profile of TCE (Figure 9) is essentially linear and does not change appreciably.   The sensitivity of this test method was much higher with Stoddard solvent.   This result suggests that the sensitivity of refractive index may be solvent-specific or contaminant-specific.

Specific Gravity.   Figure 10 shows the variation in specific gravity of TCE. Although the trend was toward a decrease in specific gravity, the sensitivity to contamination was fairly low.   The difference in specific gravity between the spent and new sample was only 0.15 percent.

Electrical Conductivity.   Figure 11 shows the electrical conductivity profile of TCE.   The electrical conductivity of a contaminated solvent was generally lower than that of a new solvent.   The same trend was observed in this case.   One particular sample (no. 3) showed an exceptionally low conductivity.   This result could be due to the sample's exposure to water or other contaminants while it was being obtained.

Sensitivity of the conductivity test was marked and consistent.   However, this test may require some form of temperature control because of the high temperature dependency of conductivity.   The decrease in conductivity was some 22 percent over the entire range of days that the solvent was used.

Visible Absorbence Spectrometry.   Visible absorbence variations are presented in Figure 12 for four wavelengths (400, 450, 500, and 600 nm).   The discoloration of TCE, as indicated by increased absorbence, increased with usage time in a linear manner.   The absorbence slope increased with decreasing wavelength.   The increase in absorbence over the entire usage period at 450 nm was about 50 percent.

Results similar to those in Figure 12 were also observed for Stoddard solvent (see Volume I).   Thus, visible colorimetry is a viable method for predicting TCE change. Rugged colorimeters that include an optical probe are available in the market.   The probe is dipped directly into any size container (e.g., test tube, vat) or can be installed permanently into a pipeline for process monitoring.   The readout (digital or analog) is instantaneous.

Paper Chromatography.   Table 5 shows the R_f values of Disperse Yellow 9 and Brilliant Oil Blue BMS dyes for TCE samples.   The results show that the solvent was still very potent at the end of the sampling cycle.   In addition, the results were contaminant-specific, i.e., trace amounts of aromatic contaminants significantly increased the R_f values although the KBV did not reflect such an increase.   The cause of this anomaly is not clear.

---

[31]G. J. Shugar, et al.; *Abbe-56 Refractometer Manual.*
[32]*Handbook of Chemistry and Physics*, 57th ed. (CRC Press, 1976).

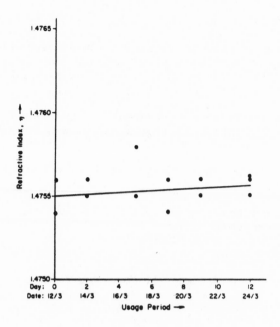

Figure 9. Variation in refractive index of trichloroethylene with usage time.

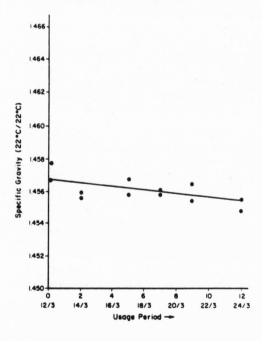

Figure 10. Variation in specific gravity of trichloroethylene with usage time.

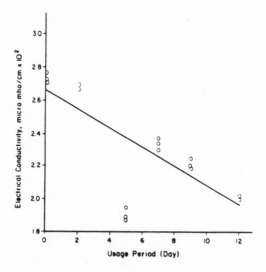

Figure 11. Variation in electrical conductivity of trichloroethylene with usage time.

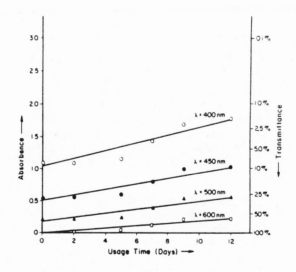

Figure 12. Variation in visible absorbence of trichloroethylene with usage time.

Table 5

R_f Values of Different Dyes on Glass Microfiber Sheet

| TCE (Day) | Dye | | | | | |
| | Brilliant Oil Blue BMS | | | | Disperse Yellow 9 | KBV |
	Violet	Blue-1	Blue-2	Blue-3		
0	1.00	1.00	1.00	1.00	0.88	129.4
2	1.00	0.84	0.72	0.57	0.38	128.3
5	0.96	0.80	0.68	0.54	0.36	127.9
7	1.00	0.85	0.74	0.60	0.43	127.8
9	1.00	1.00	1.00	1.00	0.94	127.0
12	1.00	0.84	0.73	0.60	0.40	126.9

Solids Content. This test was devised to determine the amount of particulate matter in the TCE samples. The procedure was to filter the TCE solvent followed by drying of the filter paper at 60 °C in an evacuated oven. The dried paper was weighed and the weight difference before and after filtration reported as the solids content. The results are shown in Figure 13. Contrary to expectations, the solids content did not show an increase with time.

AAV. Figure 14 shows that the AAV of TCE decreased with use. The decrease in AAV was about 40 percent over the entire period of solvent use. However, the AAV at the end of the cycle was 0.085 weight percent NaOH, which was substantially higher than the minimum recommended value (0.03 to 0.04 weight percent NaOH).

The AAV reflects the amount of stabilizers/inhibitors in TCE that was added to inhibit formation of acid when the solvent was heated in vapor-degreasing equipment. This test is generally performed daily, especially when the AAV drops below 0.06 weight percent NaOH.

PERC

Two series of PERC were obtained from Kelly Air Force Base (AFB), TX. The series were labeled PERC#1 and PERC#2 and had usage lives of 27 and 28 days, respectively. The tests done on PERC were the KBV, viscosity, refractive index, specific gravity, electrical conductivity, visible absorbence spectrometry, and AAV. The PERC test results are summarized below.

KBV. Figure 15 shows the variation in KBV of PERC with usage time for both series. At the outset, there was a difference of about 3 percent in KBVs between the two series. This difference diminished to less than 1 percent at the time of removal or change. Over the entire range, the variation was less than 5 percent for both series. Because of the relatively small change in KBV and the elaborate measurement procedure, this test did not appear very attractive for use in the field to monitor and predict solvent change. However, it could be used to monitor the quality of reclaimed solvent.

Viscosity. The variation in viscosity of PERC for the two series is shown in Figure 16. The increased viscosity in both cases was significant up to the final sample. The profiles are shown as regressed quadratic curves. The deviation in the final sample could be attributed to the makeup of new PERC in the degreaser due to a drop in liquid level.

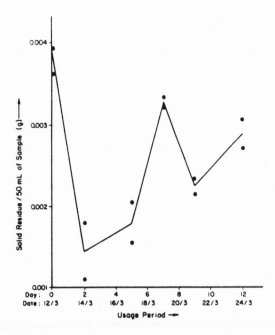

Figure 13. Variation in solids content of trichloroethylene with usage time.

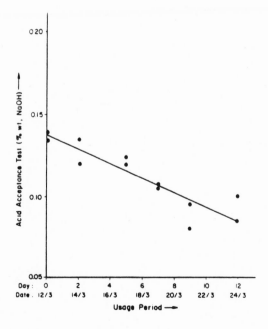

Figure 14. Variation in acid acceptance value of trichloroethylene with usage time.

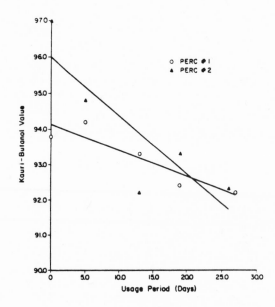

Figure 15. Variation in Kauri-butanol value of tetrachloroethylene with usage time.

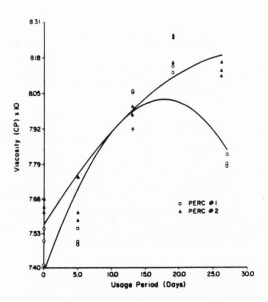

Figure 16. Variation in viscosity of tetrachloroethylene with usage time.

The solvents initially charged to the degreaser had a difference in viscosity of less than 2 percent for the two series of samples. The difference for the two series at the final sample increased to about 4 percent.

The data points oscillated appreciably, probably due to the addition of new PERC as makeup. Solvent is generally added whenever the boiling sump level falls below the level of the degreaser heating element. The makeup amount is one drum (55-gal) of new solvent.

This observation introduces a new problem in defining a cutoff point for changing solvents based on variations in a physical property such as viscosity. The profile in Figure 16 shows that the solvent in both series was not changed at the highest value of viscosity; thus, the liquid could still be used with satisfactory results. However, other properties such as color or boiling point may have changed enough to warrant solvent replacement. The solution may lie in monitoring two or more solvent properties and, when any two indicate values beyond their respective cutoff or threshold value, the solvent would be rejected. This type of criterion will become increasingly important as better methods of minimizing solvent evaporation losses are developed and implemented.

Refractive Index. Figure 17 shows the refractive index profiles of the two series of PERC samples. The refractive index decreased almost linearly with usage. The variation may depend on the type of contaminant, i.e., some contaminants may cause sharp variations whereas others may not. In this case, the contaminants were similar for both series since the refractive index indicated similar trends. This test was quite sensitive in the case of Stoddard solvent. However, TCE did not show as good a sensitivity to refractive index as PERC.

Specific Gravity. The variation in specific gravity of the two series of PERC is plotted in Figure 18. Contaminants in halogenated solvents are generally oils and grease that have a substantially lower specific gravity than the solvents. The trend is represented by a line for each of the two series. The range of decrease was roughly 1 percent.

A portable specific gravity meter is commercially available and has a range of 0 to 1.999 and an accuracy of 0.001 units. This meter was able to record the relatively small density changes encountered in all solvents tested.

Boiling Point. The boiling point trend is shown in Figure 19 for both series of PERC samples. The boiling point of pure PERC is 121.1 °C. The deviation from the boiling point of pure PERC is an approximate indication of the degree of contamination. Data for the first series of samples indicated that there was very little solvent contamination, except there was a change in color. A calibration curve for boiling point of PERC and oil mixture was obtained from Dow Chemical Company. The calibration curve indicated that PERC#2 initial sample was contaminated with the equivalent of 20 to 25 weight percent mineral oil at the outset and maintained this level of contamination until it was changed. Most solvent suppliers recommend changing solvent when the contamination level reaches 25 percent, or 125 °C in case of PERC. A high grease and oil level in the solvent reduces cleaning efficiency due to poor vapor generation and sludge formation.

Electrical Conductivity. Figure 20 shows the trend in electrical conductivity of PERC with time. Both series of samples generally show a decrease in conductivity with increased contamination. The variations between new and spent samples of PERC#1 and PERC#2 were 3.4 and 17.5 percent, respectively. The final data point for the first series deviates substantially from the observed trend, possibly because of the addition of new solvent as mentioned earlier. The sensitivity of conductivity to solvent usage time was

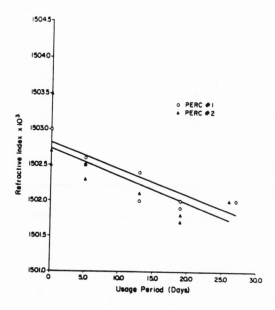

Figure 17. Variation in refractive index of tetrachloroethylene with usage time.

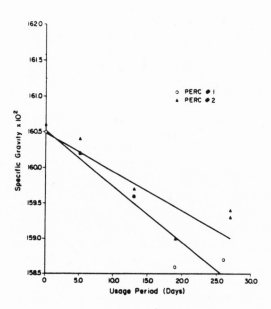

Figure 18. Variation in specific gravity of tetrachloroethylene with usage time.

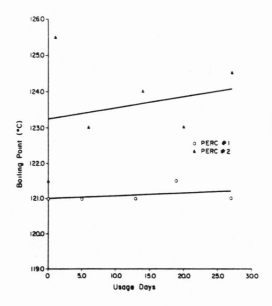

Figure 19. Variation in boiling point of tetrachloroethylene with usage time.

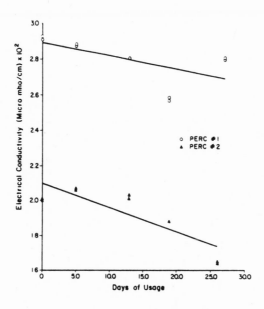

Figure 20. Variation in electrical conductivity of tetrachloroethylene with usage time.

significant and reproducible.   Some form of temperature control may be required to allow accurate evaluation of conductivity measurements.

Visible Absorbence Spectrometry. Visible absorbence variations of PERC#1 are presented in Figure 21 for four wavelengths (400, 450, 500, and 600 nm). The absorbence, indicating degree of contamination, increased with usage time in linear form. The absorbence slope increased with decreasing wavelength.   Similar results were obtained for Stoddard solvent and TCE.

The difference in absorbence between the new and totally spent solvent at 500 nm was about 391 percent. High sensitivities were also obtained at wavelengths less than 450 nm. However, the absorbence exceeded 1 at these wavelengths, i.e., the transmittance of light was less than 10 percent.

Figure 22 shows the variation in absorbence of PERC#2 with usage time. As in the case of PERC#1, this was a consistent criterion showing significant changes between different samples taken at different wavelengths (400, 450, 500, and 600 nm). The rate of change in absorbence (slope) with usage increased with decreasing wavelength. The difference between the new and spent PERC for this series was 387 percent. The final sample points showed either a decrease or no change in absorbence at all four wavelengths. This result was speculated to be due to the new solvent makeup in the degreaser.

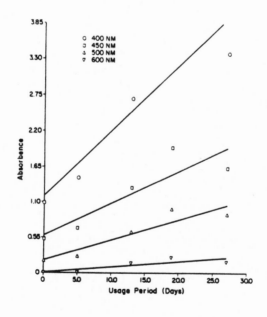

Figure 21. Variation in visible absorbence of tetrachloroethylene (PERC#1) with usage time.

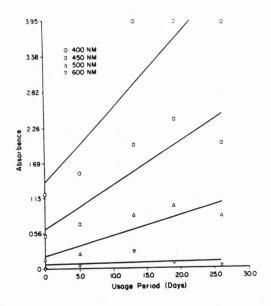

Figure 22.   Variation in visible absorbence of tetrachloroethylene (PERC#2) with usage time.

AAV.   The AAV of PERC#1 solvent varied in the following way with use:

Usage Days	AAV wt % NaOH
0	0.10
5	0.14
13	0.15
19	0.16
27	0.16

The unexpected increase in AAV, and thereby the concentration of acid-accepting inhibitor(s), in PERC#1 implied that the acid-accepting inhibitors were less volatile than PERC and that the evaporation losses of PERC were significant. This behavior was not observed with TCE. (Chapter 4 discusses the distribution of inhibitors after distillation.)

The AAV for the two series of PERC samples is graphed in Figure 23. The increase in AAV roughly corresponded to more than a 40 percent increase for PERC#1 and a 125 percent increase for PERC#2. It may be noted that values greater than 0.04 weight percent NaOH are within normal operating range; thus, the data for both series indicate no cause for concern with respect to depletion of acid-neutralizing inhibitors.

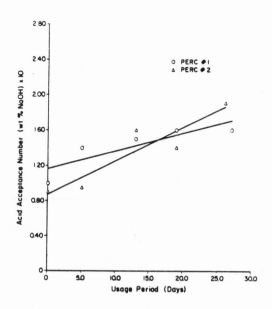

Figure 23. Variation in acid acceptance value of tetrachloroethylene with usage time.

MC

MC samples were obtained from Warner-Robins AFB, GA. The samples were taken from a vapor degreaser in an electroplating shop. The solvent in this degreaser was changed every week, irrespective of condition and performance. Samples were obtained on each of the 7 days of a solvent run. On the fourth day of this run, a drum of new MC was added because the solvent level in the degreaser fell below the steam coils. The addition of new solvent is reflected in the physicochemical tests plot as a discontinuity between the third and the fourth day (Figures 24 through 28, 30, and 31). All samples visually looked very clean and colorless. This appearance was in contrast to the yellowish-red colored spent MC samples received from Hayes International.

The tests performed on the MC samples were KBV, viscosity, refractive index, specific gravity, electrical conductivity, visible absorbence spectroscopy, AAV, and boiling point. Test results are described below.

KBV. Figure 24 shows the variation in KBV value of MC with usage time. The KBV value of the first day's sample was 118.5 and then decreased with use during the next 3 days by about 4.5 percent. After the addition of new solvent, the KBV value rose to 121.0 and then showed a slight increase until it was replaced.

The KBV of the spent solvent (6th-day sample) showed no significant change from the new solvent (1st-day sample); in fact, it increased slightly. Since the KBV indicates solvent power, this result means that the spent solvent still had the cleaning potency of new solvent.

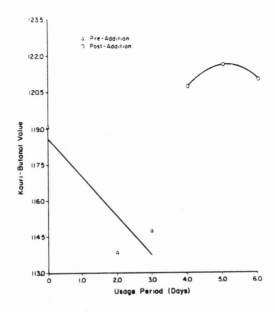

Figure 24. Variation in Kauri-butanol value of 1,1,1-trichloroethane with usage time.

Viscosity. The variation in viscosity of MC with usage time is shown in Figure 25. The initial (first sample) viscosity was about 0.79 cp and then increased to 0.80 cp prior to the addition of new solvent (fourth sample). The new solvent makeup resulted in a small decrease in viscosity of degreaser solvent to about 0.79 cp (same as that of a new solvent). There was a slight increase in viscosity when it was replaced on the sixth day. The maximum change in viscosity was less than 3 percent during the entire usage period. This relatively small change indicated that solvent contamination had taken place only to a minor extent.

Refractive Index. The variation in refractive index of MC with use is shown in Figure 26. Results indicate that no appreciable change in refractive indices occurred within the usage period. Earlier experiments with TCE and PERC had shown that refractive index was neither a very sensitive nor consistent test method.

Specific Gravity. Figure 27 plots the specific gravity profile. There was no significant change during the first 4 days of use. After makeup, the specific gravity dropped by about 0.3 percent and stayed at this value until the solvent was changed.

Electrical Conductivity. The net decrease in electrical conductivity (Figure 28) was quite marked prior to makeup with new solvent. During the first 4 days, the solvent--after an initial slight increase--decreased by about 23 percent. However, after makeup with the new solvent, the batch showed an increase in conductivity with use. This result is an anomaly compared with the tendency of most degreaser contaminants to cause a decrease in conductivity of chlorinated solvents.

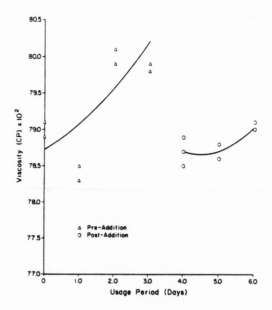

Figure 25.   Variation in viscosity of 1,1,1-trichloroethane with usage time.

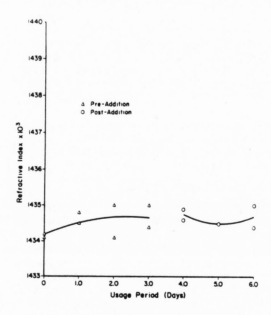

Figure 26.   Variation in refractive index of 1,1,1-trichloroethane with usage time.

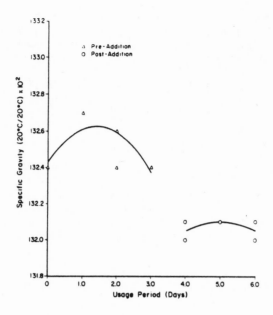

Figure 27. Variation in specific gravity of 1,1,1-trichloroethane with usage time.

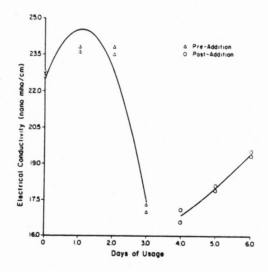

Figure 28. Variation in electrical conductivity of 1,1,1-trichloroethane with usage time.

Visible Absorbence. These measurements were done at four wavelengths (400, 450, 500, and 600 nm). Generally, lowering the wavelength increased the sensitivity of the method. However, Figure 29 shows that there was virtually no variation in absorbence at all four wavelengths. This result means that the solvent was clean and still usable.

AAV. The AAVs of the MC samples are shown in Figure 30. After a first-day increase of about 40 percent, the AAV of the solvent decreased steadily until the time of replacement. At the time of change, the AAV value was about 0.1 weight percent NaOH which is still more than twice the solvent manufacturers' recommended value for a solvent change (0.04 weight percent NaOH).

Boiling Point. The boiling point variation of MC within the 6-day sampling period was less than 0.5 °C of the first-day sample boiling point (Figure 31). This change amounts to a negligible degree of contamination because solvent manufacturers generally recommend solvent change after an increase in boiling point of about 25 °C over that of a new solvent. All test methods, except for electrical conductivity, indicated that the contamination of MC during use was negligible, making the solvent still potent for degreasing operations.

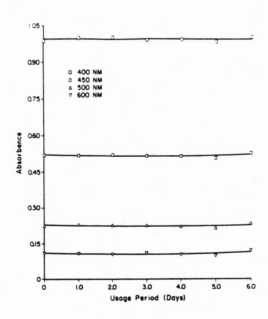

Figure 29. Variation in visible absorbence of 1,1,1-trichloroethane with usage time.

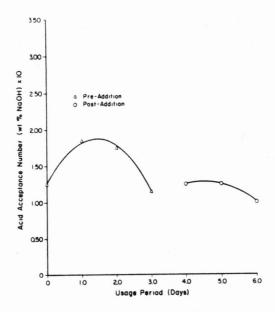

Figure 30.   Variation in acid acceptance value of 1,1,1-trichloroethane with usage time.

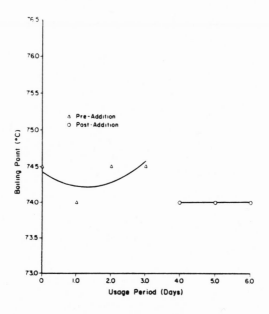

Figure 31.   Variation in boiling point of 1,1,1-trichloroethane with usage time.

Rating the Test Methods

To establish a rating, the test methods were categorized and quantified for the three chlorinated solvents on the basis of the experimental results. Factors used for the rating were:

- Sensitivity and reliability.

- Reproducibility.

- Ease of performing.

- Operator training.

- Equipment cost.

- Maintenance cost.

All factors except for sensitivity and reliability applied equally to TCE, PERC, and MC.

Sensitivity and Reliability

This factor evaluates the magnitude of change in a property with different levels of solvent contamination. The sensitivity of a method can be correlated as:

$$\% \ S = [|P_{=i} - P_{=t}| \ / \ P_{=i}] \cdot 100 \qquad\qquad [Eq\ 7]$$

where % S is percentage sensitivity, P_i is the property value of the first sample (test series 2), and P_t is the property value of the "t"th sample.

A point system was used to quantify the effectiveness of the test methods. The levels of grading for each rating factor were A, B, C, and D, or numerically 4, 3, 2, and 1 point, respectively. Sensitivity and reliability of these tests were assessed for each of the three solvents.

TCE. Figure 32 plots percentage sensitivity versus usage days for the time study samples. Visible absorption at 450 nm had a percentage sensitivity of 90.7 percent between the first and final samples. It was followed by AAV, electrical conductivity, viscosity, KBV, specific gravity, and refractive index with percentage sensitivities of 39.3, 26.4, 6.2, 2.0, 0.14, 0.007, respectively. TLC was not rated because of the uncertainties involved in measuring R_f values.

On the basis of TCE sensitivity analysis, visible absorbence and AAV were assigned an A ranking, followed by a B for electrical conductivity and viscosity, a C for KBV and specific gravity, and a D for refractive index.

PERC. Figure 33 shows percentage sensitivity versus usage days for PERC#2. PERC#1 showed a similar trend and, since two sets of data were available for PERC, the average value of the two sets was used in analyzing sensitivity.

Visible absorption at 450 nm had a percentage sensitivity of 413.2 percent between the first and final samples. It was followed by AAV, electrical conductivity, viscosity, KBV, specific gravity, boiling point, and refractive index with percentage sensitivities of

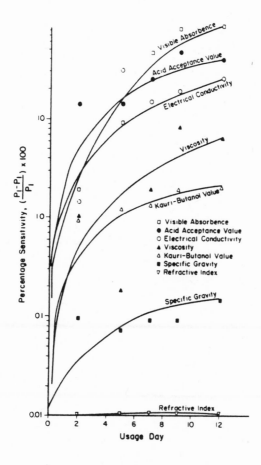

Figure 32. Sensitivity of test methods (trichloroethylene).

85.6, 11.2, 5.0, 3.33, 0.95, 0.40, and 0.05, respectively. TLC tests were abandoned because of inconsistent results with TCE.

On the basis of the PERC sensitivity analysis, the visible absorbence and AAV were assigned an A ranking, followed by a B for electrical conductivity and viscosity, a C for KBV, and a D for boiling point, specific gravity, and refractive index.

MC. Table 6 shows the calculated percentage sensitivity versus usage days for the MC time series samples. AAV was the most sensitive test method with a sensitivity of 18.4 percent, followed by electrical conductivity, visible absorbence (600 nm), KBV, boiling point, specific gravity, viscosity, and refractive index with percentage sensitivities of 12.7, 9.1, 1.62, 0.34, 0.26, 0.13, and 0.04, respectively.

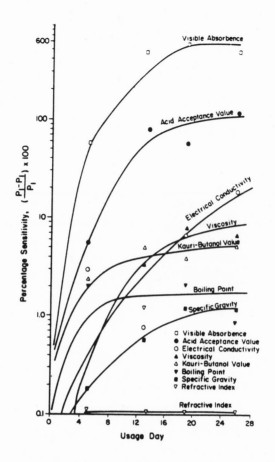

Figure 33. Sensitivity of test methods (tetrachloroethylene, PERC#2).

Table 6

Percentage Sensitivities of the Test Methods for MC

Day	AAV	Refractive Index	Specific Gravity	Electrical Conductivity	KBV	Viscosity	Visible Absorbence (600 nm)	Boiling Point (°C)
0	0.0	0.000	0.00	0.0	0.00	0.00	0.0	0.00
1	51.0	0.035	0.23	2.0	12.90	0.76	0.0	0.00
2	42.9	0.024	0.08	2.8	4.40	1.77	3.6	0.00
3	6.1	0.038	0.00	23.8	3.70	0.38	0.0	7.10
4	2.0	0.042	0.26	24.7	1.40	0.25	5.5	0.34
5	2.0	0.024	0.22	19.8	2.10	0.55	10.0	0.34
6	18.4	0.038	0.26	12.7	1.62	0.13	9.1	0.34

On the basis of MC sensitivity analysis, AAV and electrical conductivity were assigned a B ranking, followed by a C for visible absorbence, and a D for viscosity, KBV, specific gravity, boiling point, and refractive index.

Reproducibility

This factor accounts for the consistency of a method to repeatedly report the same values of a property when tested more than once.   All the methods showed good reproducibility.

Ease of Performing

This criterion rated the tests in terms of convenience, i.e., the time required to prepare reagents (if necessary) and perform the tests.   Visible absorbence, refractive index, specific gravity (by specific gravity meter), and conductivity measurements can be performed rapidly and no reagent preparations are required.   One drawback of the viscosity test is that the sample may have to be filtered prior to testing because suspended solid particles could clog the capillary of the viscometer and impede the solvent flow. The KBV measurement requires the preparation or weighing of K-B solution followed by titration with Stoddard solvent.   For the TLC using glass microfiber paper, a preheating stage is required, and dye solutions have to be prepared prior to use.   Consequently, viscosity, KBV, and TLC were ranked lower than the other tests.

Operator Training

This criterion rated the preparation and training required of an operator to perform the tests satisfactorily.   All test methods were fairly simple and did not require substantial operator training.   However, KBV and TLC required relatively more preparation and training, as well as judgment by the operator.

Equipment Cost

This factor was based on the capital cost of the apparatus required for performing each test.   Tests requiring equipment that costs $500 or less were ranked A; those costing higher than $500 but less than $1200 were ranked B; and those with a cost higher than $1200 but less than $2000 were ranked C.   Costs were obtained from either equipment suppliers' catalogs or direct vendor quotes.

The viscosity apparatus consisted of an Ostwald viscometer and a stopwatch, and the KBV apparatus was basically a buret and an Erlenmeyer flask.   This equipment cost less than $100 each and was ranked A.   A specific gravity meter and an electrical conductivity meter with probe cost around $1200 and $1000, respectively, and were ranked B in this category.   Refractometers and visible spectrometers/in-situ dipping probe colorimeters cost about $2000 each, and thus, these tests were ranked C.

Maintenance Cost

An annual operating cost was estimated for each test method, taking into account the cost of chemicals, probes, and/or supplies necessary to conduct a test at least once a week.   Maintenance cost for test equipment such as the viscometer, refractometer, spectrometer, and specific gravity meter involves buying routine cleaning supplies and should not exceed $50. Consequently, these methods were ranked A.

In contrast, KBV and conductivity determinations as well as TLC analysis require chemical and material supplies costing about $300 annually for each method, in addition to routine cleaning supplies. The KBV measurement requires periodic supplies of K-B solution. TLC requires glass microfiber paper, dyes, and chloroform. Conductivity probes may have to be replaced annually or require a platinizing kit to replenish the platinum in a used probe.

Overall Rating

The overall rating was determined by the formula shown in Table 7. Since sensitivity and reproducibility of a test are of prime importance, they were assigned weights of 4 and 2, respectively. In the case of TCE, visible absorbence was found to be the best method, followed by density, viscosity and electrical conductivity. Refractive index, KBV, and TLC are useful in specific conditions but did not satisfy all criteria of the rating.

Using the same rating formula for PERC and MC yielded the results shown in Tables 8 and 9, respectively. For PERC, AAV and visible absorbence were found to be the best methods, followed by viscosity and and electrical conductivity. In the case of MC, AAV and electrical conductivity were the best methods, followed by visible absorbence.

Table 7

Test Methods for Rating TCE

Method	Column 2 Sensitivity and Reliability	Column 3 Reproducibility	Column 4 Ease of Performance	Column 5 Operator Training	Column 6 Equipment Cost	Column 7 Maintenance Cost (Annual)	Rating**
	A* (=4, Maximum) B (=3, Good) C (=2, Fair) D (=1, Minimum)	A (=4 Excellent) through D (=1, Poor)	A (=4 Excellent) through D (=1, Poor)	A (=4 Minimum) through D (=1, Extensive)	A<$500 $500<B<$1200 $1200<C<$2000	A<$50 $200<B>50 $500<C	
Acid acceptance value (AAV)	A	A	B	B	A	A	3.8
Visible absorbence	A	A	A	A	C	A	3.8
Specific gravity	C	A	A	A	B	A	3.1
Viscosity	B	A	B	A	A	A	3.5
Electrical conductivity	B	A	A	A	B	B	3.4
Refractive index	D	A	A	A	C	A	2.6
Kauri-butanol value (KBV)	C	A	B	B	A	B	2.9

*A=4, B=3, C=2, D=1 Point.
**Rating = 1/10 (4 x Column 2 + 2 x Column 3 + Column 4 + Column 5 + Column 6 + Column 7).

Table 8

Test Methods for Rating PERC

Method	Column 2 Sensitivity and Reliability A* (=4, Maximum) B (=3, Good) C (=2, Fair) D (=1, Minimum)	Column 3 Reproducibility A (=4 Excellent) through D (=1, Poor)	Column 4 Ease of Per- formance A (=4 Excellent) through D (=1, Poor)	Column 5 Operator Training A (=4 Minimum) through D (=1, Extensive)	Column 6 Equipment Cost A<$500 $500<B<$1200 $1200<C<$2000	Column 7 Maintenance Cost (Annual) A<$50 $200>B>$50 $500<C	Rating**
Acid accep- tance value (AAV)	A	A	B	B	A	A	3.8
Visible absorbence	A	A	A	A	C	A	3.8
Specific gravity	D	A	A	A	B	A	2.7
Viscosity	B	A	B	A	A	A	3.5
Electrical conductivity	B	A	A	A	B	B	3.4
Refractive index	D	A	A	A	C	A	2.6
Kauri-butanol value (KBV)	C	A	B	B	A	B	2.9
Boiling point	D	A	B	B	A	A	2.6

*A=4, B=3, C=2, D=1 Point.
**Rating = 1/10 (4 x Column 2 + 2 x Column 3 + Column 4 + Column 5 + Column 6 + Column 7).

Evaluation of Batches of Spent Solvent

The consistency of the test methods for random batches of spent solvent is an important factor in developing a criterion for identifying a spent solvent. To achieve this objective, three batches of spent PERC taken at different times were obtained from Kelly AFB.

The three spent PERC samples were labeled PERC#1, PERC#2, and PERC#3. PERC#1 and PERC#2 had residence times of 27 and 26 days, respectively. The residence time of PERC#3 could not be determined. However, all three samples were taken between 15 March and 8 May 1986.

Table 10 shows the variation in physicochemical properties among the three PERC samples. Of all properties, electrical conductivity, AAV, and visible absorbence revealed the greatest variations (41.2, 31.6, and 23.9 percent, respectively). These three test methods, however, were the most sensitive. The high variations could be due to two factors: (1) the samples were used to different degrees, although the numbers of days in the vat may be similar and (2) the makeup of solvents in degreasers made it difficult to track the solvent condition. Variations in other solvent properties--KBV, boiling point, viscosity, refractive index, and specific gravity--were less than 5 percent.

AAV is an essential test for chlorinated solvents to verify if acid-accepting inhibitors are present in adequate levels. However, AAV cannot determine the potency of a solvent because it is simply a hydrochloric acid titration procedure. Therefore, it has to be supplemented by one or more tests that indicate solvent quality (e.g., visible absorbence, conductivity).

The physical properties of a solvent are temperature-dependent. A variation in ambient temperature is generally avoided by using a water bath. However, in the field, the use of a water bath apparatus would seriously impede easy handling of the test equipment and reduce mobility. This problem could be resolved by testing the new solvent along with the used product. The new solvent value would serve as a ratio or

Table 9

Test Methods for Rating MC

Method	Column 2 Sensitivity and Reliability A* (=4, Maximum) B (=3, Good) C (=2, Fair) D (=1, Minimum)	Column 3 Reproducibility A (=4 Excellent) through D (=1, Poor)	Column 4 Ease of Per-formance A (=4 Excellent) through D (=1, Poor)	Column 5 Operator Training A (=4 Minimum) through D (=1, Extensive)	Column 6 Equipment Cost A<$500 $500<B<$1200 $1200<C<$2000	Column 7 Maintenance Cost (Annual) A<$50 $200<B>50 $500<C	Rating**
Acid acceptance value (AAV)	B	A	B	B	A	A	3.8
Visible absorbence	C	A	A	A	C	A	3.0
Specific gravity	D	A	A	A	B	A	2.7
Viscosity	D	A	B	A	A	A	2.7
Electrical conductivity	B	A	A	A	B	B	3.4
Refractive index	D	A	A	A	C	A	2.6
Kauri-butanol value (KBV)	D	A	B	B	A	B	2.5
Boiling point	D	A	B	B	A	A	2.6

*A=4, B=3, C=2, D=1 Point.
**Rating = 1/10 (4 x Column 2 + 2 x Column 3 + Column 4 + Column 5 + Column 6 + Column 7).

Table 10

Percentage Consistency of Spent Solvent (PERC)

Test Methods	New PERC	PERC #1 (Used 27 days)	PERC #2 (Used 26 Days)	PERC #3 (Usage Time Unknown)	Maximum Difference
KBV	94.6000	92.2000	92.2000	----	0
Viscosity (cp)	0.7500	0.7800	0.8100	0.8000	4.0600
Refractive index	1.5030	1.5020	1.5020	1.5019	0.0070
Specific gravity	1.6070	1.5940	1.5870	1.5770	1.0600
Visible Absorbence (500 nm)	0.1830	0.9000	0.8290	1.0900	23.9000
Boiling point (°C)	122.0000	121.0000	124.5000	122.0000	2.0000
Electrical conductivity (nanomho/cm)	29.4000	27.9000	16.4000	27.0000	41.2000
AAV (% wt, NaOH)	0.1000	0.1600	0.1900	0.1300	31.6000

nondimensionalizing factor in determining the criterion. The nondimensional quantities for the spent solvent are obtained by the following relationship:

$$NDP = | C_m - C_o | / C_o \qquad\qquad [Eq\ 8]$$

where: NDP = nondimensional property under study.
C_m = measured property value of spent solvent.
C_o = measured property value of new (unused) solvent.

The discussion of the test methods for Stoddard solvent (Volume I) indicated that the use of multiple test methods increases the accuracy of characterizing spent solvent. The discussion in this chapter also suggests that using multiple criteria (test methods) instead of just one to monitor a solvent increases the reliability of the solvent change process. A simple, quantitative determination for multiple criteria to signal a change of solvent is to use a rating system for each criterion.   The priority of the criteria established in Tables 7 through 9 allows the following points to be assigned:

- 4 points to meet the absorbence cutoff limit.

- 3 points to meet the conductivity cutoff limit.

- 3 points to satisfy the viscosity cutoff limit.

- 2 points to meet the specific gravity cutoff limit.

A solvent is changed whenever the AAV cutoff limit is reached or 6 points or more are accumulated by other test criteria. Tables 11 through 13 list the cutoff limits for the three chlorinated solvents obtained in the experimental study. Since viscosity measurements did not indicate an appreciable variation within MC samples, as is evident from Table 6, it was excluded from Table 13.

Summary of Findings

The findings discussed in this chapter can be summarized as follows:

1. Laboratory tests showed that some physicochemical and electrical properties of chlorinated solvents such as TCE, PERC, and MC can be used successfully as criteria to identify spent solvents.

2. AAV and visible absorbence were rated as the best criteria for TCE. Viscosity and electrical conductivity were also determined to be effective criteria. KBV, specific gravity, boiling point, TLC, and refractive index methods did not satisfy all requirements for reliable criteria, but may be useful in specific situations.

3. For PERC, the AAV and visible absorbence tests were the best rated criteria, followed by viscosity and electrical conductivity.

4. In the case of MC, the AAV and electrical conductivity were the best criteria, followed by absorbence.

5. These criteria were tested over several batches of spent PERC and were fairly consistent. The identification of spent solvent, however, is most accurate when at least two or.more criteria are employed. A single criterion may occasionally lead to erroneous conclusions, resulting in either premature removal of solvent or unsatisfactory cleaning performance through prolonged use.

Table 11

Test Criteria for Trichloroethylene

Rating	Acid Acceptance Value (wt % NaOH)	Absorbence (450 nm)	Viscosity (cp)	Conductivity (nanomho/cm)
0	>0.06	<0.50	0.57	>27.0
1	--	0.50-0.67	0.571-0.59	27.0-24.0
2	--	0.68-0.84	0.591-0.60	23.9-20.0
3	--	0.85-1.00	>0.60	<20.0
4	0.06-0.03	>1.00	--	--
6	<0.03	--	--	--

Table 12

Test Criteria for Tetrachloroethylene

Rating	Acid Accept-ance Value (wt % NaOH)	Absorbence (500 nm)	Viscosity (cp)	Conductivity (nanomho/cm)
0	>0.06	<0.18	0.75	>29.4
1	--	0.18-0.42	0.76-0.77	29.4-26.7
2	--	0.43-0.66	0.78-0.80	26.6-24.0
3	--	0.67-0.90	>0.80	<24.0
4	0.06-0.03	>0.90	--	--
6	<0.03	--	--	--

Table 13

Test Criteria for 1,1,1-Trichloroethane

Rating	Acid Accept-ance Value (wt % NaOH)	Absorbence (400 nm)	Conductivity (nanomho/cm)
0	>0.06	<0.98	>22.7
1	--	0.98-0.986	22.7-21.1
2	--	0.987-0.994	21.0-19.5
3	--	0.995-1.00	<19.5
4	0.06-0.03	>1.00	--
6	<0.03	--	--

4. Chlorinated Solvent Inhibitors

Overview

Large amounts of chlorinated solvents are currently used in metal cleaning operations (vapor degreasing and cold cleaning). Consequently, huge quantities of waste solvent are generated regularly by these cleaning operations. The high cost of new chlorinated solvent combined with that of hazardous waste disposal make it necessary to control the amount of wastes produced as closely as possible. At present, the criteria for disposal of these solvents are rather arbitrary. Therefore, a scientific basis for determining solvent quality is needed to achieve the most efficient solvent use.

One major criterion for determining the quality of chlorinated solvents is the inhibitor level. Inhibitors are present in chlorinated solvents to prevent: (1) solvent breakdown, (2) the solvent's becoming an acid, and (3) solvent-part reactions. Adequate solvent inhibitor levels are therefore extremely important for safe, efficient operation. Determining a solvent's inhibitor levels is critical to the determination of solvent quality.

To demonstrate the role of solvent inhibitors in cleaning operations and investigate the potential for reclaiming waste solvent, this chapter:

1. Identifies and analyzes inhibitors.

2. Determines relationships of inhibitor concentration versus time in a typical degreasing operation.

3. Assesses inhibitor loss in solvent reclamation and other reactions.

4. Determines the relationship between inhibitor concentration and solvent life.

Literature Review

Chlorinated solvents are widely used in industrial metal cleaning as cold-cleaning and vapor degreasing agents. They are especially suited for vapor degreasing because their high vapor density (3 to 7 times heavier than air) makes them relatively easy to control as a vapor zone within the degreaser by use of a simple condenser. Also, their high solubility for oils and resistance to flashing make them extremely attractive for this type of metal cleaning operation. They are equally well suited for cold-cleaning operations when adequate precautions are taken to protect workers from the toxicity hazards of chlorinated solvents.[33]

The three most common chlorinated solvents used in metal cleaning are TCE, PERC, and MC. Table 14 lists the physical properties and process data for these solvents.

---

[33]USEPA, August 1979; T. J. Kearney and C. E. Kircher, April 1960 and May 1960; R. Monahan; R. L. Marinello.

Table 14

Physical Properties and Process Data for Chlorinated Solvents Used in Vapor Degreasing*

Solvent Properties and Data	Tetrachloro-ethylene	1,1,1-Trichloro-ethane	Trichloro-ethylene
Chemical formula	C_2Cl_4	$C_2H_3Cl_3$	C_2HCl_3
Boiling point, °F	250	165	188
Latent heat of vaporization (boiling point), Btu/lb	90	102	103
Specific heat (liquid), Btu/lb, °F	0.21	0.25	0.23
Specific gravity Vapor (air = 1.00) Liquid (water = 1.00)	5.73 1.62	4.55 1.33	4.54 1.46
Liquid density, lb/gal at 77°F (25°C)	13.5	11.0	12.1
Vapor density at boiling point, lb/cu ft	0.326	0.279	0.278
Freezing point, °F	-8	-34	-123
Steam pressure to boil, psig	50 - 60	1 - 6	5 - 15
Hot water temperature to boil, °F (pressure psig)	300 - 325 (70 - 105)	230 - 270 (20 - 50)	250 - 300 (25 - 70)
Safety vapor control thermostat setting, °F	220	130	160
High temperature cutoff thermostat (boil chamber), °F	295	190	240
Maximum temperature for effective water separator functioning, °F	190	149	164
Maximum boiling point of contaminated solvent before cleaning, °F	260	175	195
Molecular weight	165.85	133.42	131.40
Vapor pressure at 77°F (25°C), mm Hg	18	132	80
Saturated vapor concentration at 77°F (25°C), ppm	24,300	173,700	105,000

*Source: R. Monahan, "Vapor Degreasing With Chlorinated Solvents," *Metal Finishing* (November 1977). Used with permission.

Table 14 (Cont'd)

Solvent Properties and Data	Tetrachloro-ethylene	1,1,1-Trichloro-ethane	Trichloro-ethylene
Liquid volume (mL) which, if evaporated in static air in an enclosure of 1000 cu ft, would give a vapor concentration equal to time-weighted average	11.7	40.2	10.1
Volume of air (cu ft) necessary to dilute vapor from one point to concentration equal to time-weighted average	24,000	8600	30,500
Threshold limit values established by American Conference of Governmental Industrial Hygienists, volume ppm (mg/m^3)	100 (670)	350 (1900)	100 (535)
American National Standard Institute, Inc., standards for: 8-hr time-weighted average, volume ppm	100	400	100
Acceptable ceiling, volume ppm	200	500	200
Acceptable maximum peak above the acceptable ceiling concentration for an 8-hr shift, volume ppm	300 (5 min in any 3 hr)	800 (5 min in any 2 hr)	300 (5 min in any 2 hr)
Flash point, °F	none	none	none
Autoignition temperature, °F	none	856	780
Vapor flammability limits in air atmospheric pressure, volume % At 77 °F	none	6.8-10.5 10.0-15.5	8.0-10.5 none
At 212 °F	none	6.3-15.5	10.5-41 11-38
Underwriters Laboratories flammability rating scale	0	5-10	3
Odor slight, not unpleasant, volume ppm	150	350	200
Odor strong, unpleasant, volume ppm	400	1500	600
Human response to solvent vapors: No response, volume ppm	100, 8 hr	500, 7 hr daily, 5 days	100, 8 hr daily, 5 days; 200, 3 hr
Eye irritation, volume ppm	400	1000	400 slight 1000 definite
Respiratory irritation, volume ppm	600	2000	1000

Table 14 (Cont'd)

Solvent Properties and Data	1,1,1-Tetrachloro-ethane	Trichloro-ethylene	Trichloro-ethylene
Minimal anesthesia (light-headed, dizzy feeling), volume ppm	200, 8 hr; 400, 2 hr; 600, 10 min	1000, 30-70 min. 1500, 15-60 min; 2000, 5 min	400, 20 min; 1000, 6 min; 1500 < 5 min
Kauri-butanol value	90	125	130
Viscosity at 77 °F (25 °C)	0.84	0.79	0.54
Dielectric constant at 68 °F (20 °C)	2.30	7.1	3.27
Surface tension at 68 ° (20 °C), dynes/cm	32.3	25.9	32.0
Evaporation rate (CCl_4 = 1.0)	0.25	0.9	0.8
Solubility Solvent in water at 77 °F (25 °C), weight %	0.02	0.07	0.1
Solvent in acetone, benzene, ethyl, ether, n-heptane and methanol	miscible	miscible	miscible
Pounds per gallon at 77 °F (25 °C)	13.47	10.98	12.12

Solvent Inhibitors

Chlorinated solvents contain three basic types of inhibitors: antioxidants, acid acceptors, and metal reaction stabilizers. The combination of inhibitors added to the solvent depends on the characteristics of the solvent and the cleaning operation.

Antioxidants. Unsaturated chlorinated solvents (TCE and PERC) are prone to autoxidation. Lundberg has proposed a free radical chain mechanism for this type of oxidation of unsaturated compounds.[34] First, an initiator (heat, ultraviolet, etc.) causes the removal of a hydrogen atom from an unsaturated molecule, RH, forming a free radical (R). A molecule of oxygen (O_2) combines with this free radical to form a peroxy radical, ROO · , which can then remove a hydrogen atom from a new unsaturated

[34] W. O. Lundberg, *Symposium on Foods: Lipids and Their Oxidation*, H. W. Schultz, et al. (Eds.) (AVI Publishing, 1962).

molecule, R'H, propagating the chain reaction.   Lundberg's proposed free radical
mechanism is:

Initiation

$$RH \text{ -----------------------> } R^{\cdot} + H^{\cdot}$$

Unsaturated hydrocarbon   heat, metals, light, enzyme
                                 $R^{\cdot} + H^{\cdot}$
                                 alkyl free radical

$$\text{or } RH + O_2 \text{ ---------------> } R^{\cdot} + HO_2^{\cdot}$$

$$R^{\cdot} + O_2 \text{ <----------------- } ROO^{\cdot}$$
                                 peroxy radical

Propagation

$$ROO^{\cdot} + R'H \text{ ---------------> } ROOH + R'^{\cdot}$$
                                 hydroperoxide

Decomposition

$$ROOH \text{ --------------------> } RO^{\cdot} + HO^{\cdot}$$
                                 alkoxy free radical

$$RO^{\cdot} + RH \text{ ----------------> } ROH + R^{\cdot}$$

$$HO^{\cdot} + RH \text{ ----------------> } H_2O + R^{\cdot}$$

$$ROOH \text{ --------------------> } ROO^{\cdot} + H^{\cdot}$$

$$ROOH \text{ --------------------> } R^{\cdot} + HO_2^{\cdot}$$

Termination

$$R^{\cdot} + R^{\cdot} \text{ ------------------> } RR$$

$$R^{\cdot} + H^{\cdot} \text{ -----------------> } RH$$

$$RO^{\cdot} + HO^{\cdot} \text{ ---------------> } ROOH$$

$$ROO^{\cdot} + H^{\cdot} \text{ ---------------> } ROOH$$

$$RO^{\cdot} + R^{\cdot} \text{ ----------------> } ROR$$

$$R^{\cdot} + ROO^{\cdot} \text{ ---------------> } ROOR$$

$$ROO^{\cdot} + ROO^{\cdot} \text{ -------------> } \text{Unknown}$$

Antioxidants can be classified into three groups: phenols, amines, and amino-
phenols.   All of these compounds contain an unsaturated benzene ring with a phenol
and/or an amine group. They are effective antioxidants because of their ability to form
stable resonance hybrids after losing a hydrogen atom to an oxidation free radical as

shown in Figure 34, thus slowing the propagation step of the autoxidation.[35]   Some antioxidants also have an effect by reacting with or decomposing the hydroperoxide intermediates.

Metal Stabilizers.   MC is essentially stable to autoxidation; however, it is prone to degradation in the presence of aluminum or aluminum chloride ($AlCl_3$).   TCE and PERC are also somewhat prone to degradation in the presence of $AlCl_3$ or aluminum fines.[36] These mechanisms are shown in Figures 35 and 36.   To prevent these reactions, Lewis base metal stabilizers are added to chlorinated solvents.   These stabilizers may function by complexing or reacting with $AlCl_3$ or by terminating free radicals through hydrogen donation.   Archer and Simpson state, however, that the main function of the stabilizer is to compete with the solvent for electron-deficient sites on the chemiadsorbed $AlCl_3$ on the metal or metal oxide surface.[37]   The stabilizer converts the $AlCl_3$ into an insoluble coating on the metal surface:

Active Al Site + Inhibitor ---> Al-Inhibitor Complex

$AlCl_3$ (Chemiadsorbed) + Inhibitor ---> Insoluble Surface Coating

This proposed reaction was substantiated by the observation of insoluble deposits at microreaction sites on the metal surface of an aluminum coupon that had been placed under long-term contact with boiling commercial MC.   Examination of these deposits using an electron probe showed them to be high in carbon and chlorine content.[38]

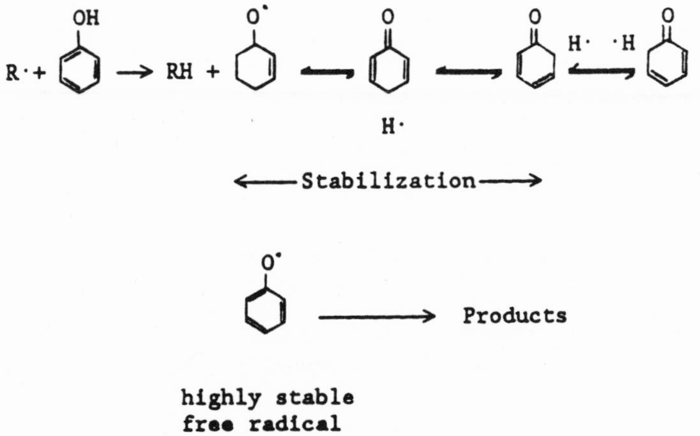

Figure 34.   Resonance hybridization of antioxidants.   (Source: A. S. Pereira, *Composition and Stability of Poultry Fats*, Ph.D. Thesis [Purdue University, 1975].)

---

[35]A. S. Pereira, *Composition and Stability of Poultry Fats*, Ph.D. Thesis (Purdue University, 1975).
[36]W. L. Archer and E. L. Simpson, *Ind. Eng. Chem., Prod. Res. Dev.*, Vol 16, No. 2 (1977), pp 158-162; W. L. Archer, *Ind. Eng. Chem., Prod. Res. Dev.*, Vol 21 (1982), pp 670-672.
[37]W. L. Archer and E. L. Simpson.
[38]W. L Archer.

Chloride Induced Breakdown of the Aluminum Oxide Film

$$Al(OH)_3 \rightleftharpoons Al(OH)_2^+ + OH^-$$

$$Al(OH)_2^+ + Cl^- \longrightarrow Al(OH)_2 2Cl$$
$$\text{* soluble aluminum}$$
$$\text{hydroxychloride salt}$$

* The soluble salt is removed from the oxide structure by complexing
with the solvent. The vacant Al orbital is satisfied by the donation
of an electron from a Cl atom in the solvent to the Al atom.

Reaction Sequence Between Aluminum and MC

$$[\gg Al^{3+} \gg Al^3 \;]\text{-}Cl^-_{ads} + 3[Cl\text{-}\overset{Cl}{\underset{Cl}{C}}\text{-}CH_3] \longrightarrow$$

$$[\gg Al^{3+}]Al^{3+} + 4Cl^- + 3(CH_3Cl_2C\cdot)$$

$$Al_{surface} + 3(CH_3Cl_2C\cdot) \longrightarrow 3/2(CH_3CCl_2CCl_2CH_3)$$

Figure 35. Proposed mechanism for the aluminum-MC degradation reaction.
(Source: W. L. Archer, *Ind. Eng. Chem., Prod. Res. Dev.*, Vol 21
[1982]. Used with permission.)

MC

$$CH_3CCl3 \xrightarrow{AlCl_3} CH_2=CCl_2 + HCl$$
$$\text{dehydrohalogenation}$$

TCE

$$Cl_2C=CHCl \xrightarrow{AlCl_3} Cl_2CHCCl_2CH=CCl_2$$

$$Cl_2CHCCl_2CH=CCl_2 \longrightarrow CCl_2=CClCH=CCl_2 + HCl$$
$$\text{dimerization}$$

Figure 36. Degradation reactions of chlorinated solvents in the presence of
$AlCl_3$. (Source: W. L. Archer, *Ind. Eng. Chem., Prod. Res. Dev.*, Vol
21 [1982]. Used with permission.)

Van Gemert[39] believes that for sufficient stabilization to occur, inhibitors must destroy $AlCl_3$ as well as prevent its formation in the reaction stated by Archer. Commercial stabilizer concentrations are less than 5 percent and cannot prevent the formation of $AlCl_3$ but can only slow it. Therefore, it is necessary to include compounds that destroy $AlCl_3$ (alcohols, epoxides) and prevent solvent decomposition.[40]

Acid Acceptors. Each of these three solvents (TCE, PERC, and MC) break down to form small amounts of HCl. This HCl can cause several problems in operation--including contributing to solvent degradation and corrosion of parts and the degreaser system.

Therefore, acid acceptance inhibitors are added to these solvents to "scavenge" molecules of HCl. These acid acceptors are generally amines or epoxides, which will react with the small amounts of HCl formed by the degradation of chlorinated solvents and neutralize them. An example is the reaction of an epoxide with HCl as follows:[41]

$$\underset{CH_2 \longrightarrow CH_2}{\overset{O}{\triangle}} + HCl \longrightarrow \underset{CH_2 \longrightarrow CH_2}{\overset{OH \quad Cl}{|\qquad |}}$$

Inhibitor Packages

As mentioned earlier, a combination of inhibitors is added to a solvent (i.e., an inhibitor package). This package may consist of compounds from any or all of the three inhibitor classes. The exact makeup of the package depends on the characteristics of the solvent and of the cleaning operation (i.e., soil, metal involved). These inhibitor packages are designed to achieve the safest, most efficient operation possible. In general, TCE and PERC packages are mainly antioxidants and acid acceptors (inhibitor concentration < 1 percent by weight), whereas MC packages contain mainly metal stabilizers and acid acceptors (total inhibitor concentration 4 to 6 percent by weight).[42]

Experimental Procedures

Sampling

Samples of vapor degreasing solvents were obtained from typical DOD metal cleaning installations. Table 4 (Chapter 3) summarized installations supplying solvent samples for testing.

Identification of Solvent Inhibitors

Inhibitors were identified using gas-liquid chromatography (GLC). GLC is an extremely versatile tool for analyzing volatile solvents. This technique separates sample components by partitioning solutes between a mobile gas phase and a stationary liquid phase held on a solid support. A sample is injected into a heating block where it is vaporized, and the resulting vapor plug is carried to the column inlet by the carrier gas. The

[39]B. Van Gemert, *Ind. Eng. Chem., Prod. Res. Dev.*, Vol 21 (1982), pp 296-297.
[40]B. Van Gemert.
[41]C. D. Gutsche and D. J. Pasto, *Fundamentals of Organic Chemistry* (Prentice-Hall, 1975).
[42]W. L. Archer; B. Van Gemert.

solutes are adsorbed at the head of the column by the stationary phase and then desorbed by the carrier gas. This adsorption-desorption process occurs repeatedly as the sample moves down the column, with each component traveling at its own rate. The components will separate to a degree determined by their partition ratios and their band spreading. The solutes will elute sequentially with increasing value of their partition ratios and enter a detector at the end of the column. Some common detectors are the thermal conductivity detector (TCD), the flame ionization detector (FID), the flame photometric detector (FPD), and the electron capture detector (ECD).[43]

Gas chromatography using a mass spectrometer as the detector (GC-MS) was used for primary identification of inhibitors in new TCE, PERC, and MC. The system used for GC-MS was a Varian 3700 GC coupled to a VG E-HF Magnetic Sector mass spectrometer.

A mass spectrometer ionizes gaseous molecules, separates the ions produced on the basis of mass-to-charge ratio (m/e) and then records the relative number of different ions produced. The m/e is then plotted as the abscissa with relative intensity as the ordinate. This plot is referred to as a "mass spectrum." The mass spectrum of a compound can be considered its "fingerprint" and can be used to identify a compound through comparison with published reference spectra. Most mass spectrometers in use today are interfaced with computers that can compare experimental spectra to the standards and thus perform the identification automatically. Sometimes, however, an experimental spectrum will not closely match any of the published spectra and will require the experimenter to make the identification. To perform this type of identification, the molecular weight of the compound must be determined or estimated. This is done using the molecular peak of an electron-impact ionization spectrum or can be measured more accurately using chemical ionization MS. The experimental spectrum is then compared with spectra of compounds of the same molecular weight in a computerized spectra database or a published library of spectra.[44]

The results of the GC-MS analysis were then further confirmed, when possible, through GC retention time studies (also referred to as "spiking"). The GC was a Varian 3700 equipped with a J&W DB-5 capillary column and FID. The integrator was a Hewlett Packard 3390A.

The FID is a very versatile detector for analysis of organic compounds. Column effluent enters a burner base through a millipore filter, is mixed with hydrogen gas, and is then burned at a flame tip with air or oxygen. This process forms ions and free electrons. These enter a gap between two electrodes (applied potential approximately 400 V), the flame jet, and a collector, thus causing the current to flow. This current flow, sensed as a voltage drop, is amplified and can be displayed on a recorder. Therefore, FID responds only to substances that produce ions when burned in a hydrogen/air or hydrogen/oxygen flame. In an organic compound, when $-CH_2-$ groups are introduced into the flame, positively charged carbon groups and electrons are formed, causing a large increase in current. The response is proportional to the number of oxidizable carbon atoms.[45]

Spiking is an extremely simple and useful method for the further confirmation of a compound's identity. A small amount of pure compound is added to the sample to be

---

[43]H. H. Willard, L. L. Merritt, Jr., J. A. Dean, and F. A. Settle, Jr., *Instrumental Methods of Analysis*, 6th ed. (Wadsworth Publishing, 1981).

[44]H. H. Willard, et al.

[45]H. H. Willard, et al.

analyzed. This sample is then injected into the GC. The resulting chromatogram is then inspected for the presence of new peaks. If the unknown and pure compounds are the same, there should be no new peaks on the "spiked" chromatogram (as compared with the original chromatogram), with the only difference in the chromatograms being in the relative area of the peak in question. "Spiking" can be used with any type of detector (TCD, FID, FPD, or ECD).

Solvent Inhibitor Time Studies

Samples of vapor degreasing solvents taken at set intervals during the vapor degreasing cycle (new solvent to spent solvent) were analyzed for inhibitor concentration using the internal standard method and GC with an FID. The internal standard method is commonly used in GC as a method of measuring sample component concentration. It can be used with a TCD, FID, ECD, or FPD. A known amount of pure substance (the internal standard) is added to standard solutions and samples to achieve a known concentration of the internal standard. The standard solutions are then used to construct a calibration curve (response ratio vs. analyte concentration), which can be used to determine the analyte concentrations of the samples.[46] These data can be used to find the relationship between solvent inhibitor concentration and time.

The concentration versus time study can also be compared with the acid acceptance versus time study. AAV (ASTM D 2942) is the most common method of determining the inhibitor level of chlorinated solvents.[47] To be specific, it measures the acid acceptance inhibitor level present in these solvents.

The total AAV is determined by reaction with nonaqueous hydrochloric acid (HCl) in excess. The excess acid is then neutralized with a standard sodium hydroxide (NaOH) solution.

The procedure is as follows:

1. Blank: pipet 25 mL of hydrochlorinating agent (0.1 N HCl in isopropanol) and 10 mL of isopropanol into a 250-mL Erlenmeyer flask. Add 3 drops of bromophenol blue indicator solution (1 g bromophenol blue in 800 mL water and 200 mL denatured ethanol) and titrate to a stable endpoint with 0.1 N NaOH solution.

2. Sample: pipet 25 mL of hydrochlorinating solution into a glass-stoppered Erlenmeyer flask. Add 10 mL of chlorinated (halogenated) solvent and 25 mL of isopropanol. Shake thoroughly and allow to stand at room temperature for 10 min. Add 3 drops of bromophenol blue indicator solution to the flask and titrate to a stable endpoint with 0.1 N NaOH solution. The AAV can now be calculated as weight percent NaOH:

$$AAV \text{ (wt \% NaOH)} = [(A-B)N \cdot 0.04 \cdot 100]/W \qquad \text{[Eq 9]}$$

where: A = mL NaOH solution required for titration of the blank.
              B = mL NaOH solution required for titration of the sample.
              N = normality of the NaOH solution.
              W = grams of solvent sample used.

---

[46]H. H. Willard, et al.
[47]*Annual Book of ASTM Standards.*

The constants 0.04 and 100 are the conversion factors from (mL·equivalent/L) NaOH to g NaOH, and from weight fraction Na to weight percent NaOH.

The inhibitor concentrations can be compared with the AAVs of the samples to determine how acid acceptance levels correspond with total inhibitor levels (i.e., does acid acceptance accurately determine if levels of metal stabilizers or antioxidants are adequate). In short, it will show whether acid acceptance is a good test for monitoring the total inhibitor level of a solvent. Also, it can be used to help determine which inhibitors act as acid acceptors.

Solvent Reclamation Inhibitor Studies

Studies were performed to determine if there are any changes in solvent inhibitor concentration in reclaimed solvents compared with (1) used and (2) new solvent. Used solvents were reclaimed using distillation and activated carbon adsorption. Inhibitor concentrations were again determined using the internal standard method and GC.

Distillation. A bench-scale distillation apparatus was used to determine the recoverability of the inhibitors in solvent reclamation by distillation. Approximately 50 mL of used solvent was pipetted into the apparatus. The distillation was done at atmospheric pressure in glass, using an electric heating mantle as a heat source. The heat source was removed when boiling had ceased, and there was a significant change in the boiling temperature. The distillate was then analyzed for inhibitor concentrations. These values were compared with the inhibitor concentrations for new and spent (solvent before reclamation) solvent. This process was performed for TCE, PERC, and MC.

Activated Carbon Adsorption. About 50 g of used solvent and 5 g of activated carbon (Darco grade HDC, ICI Americas, Inc.) were added to a 125-mL Erlenmeyer flask equipped with a magnetic stirring bar. The flask was then stoppered and allowed to remain overnight at room temperature with constant stirring. The mixture was filtered using a Buchner funnel and filter paper and the filtrate was analyzed for inhibitor levels. Again, these results were compared with the inhibitor concentrations of new and spent solvent. This process was performed for all of the three solvents.

Inhibitor Kinetic Studies

Batch Reactions. Batch reactor kinetic studies were performed to determine the reaction orders and rate constants of several acid acceptor/HCl reactions and metal stabilizer/$AlCl_3$ reactions. These reactions were performed using uninhibited PERC (HPLC grade, 99.9 percent, Aldrich Chemical Co.) as solvent. It was chosen because of its stability.

The acid acceptor/HCl reactions were performed as follows:

1. A measured amount of a PERC solution (9.5 g) with a certain concentration of acid acceptor (2.7×10^{-3} mole/L) and internal standard (n-decane, 99+ percent, Sigma Chemical Co.) were placed into a reaction vessel (a glass vial).

2. The reaction vessel was placed into a constant-temperature water bath and the solution was allowed to reach the bath temperature.

3. The required amount of a nonaqueous HCl solution, 5.2×10^{-4} mole/L HCl in isopropanol (2-propanol, Fisher certified ACS, Fisher Scientific), to completely react

with the acid acceptor (0.5 g) was measured into a syringe which was then placed into the water bath and allowed to reach the bath temperature. The stoichiometry of the reaction was found in the literature and was verified experimentally.

4. The HCl solution was injected into the reaction vessel as a slug. The concentration of the acid acceptor was monitored using GC and the internal standard method.

The acid acceptors studied were butylene oxide (1,2-epoxybutane, gold label, 99.5%, Aldrich Chemical), cyclohexene oxide (98 percent, Aldrich), and epichlorohydrin (gold label, 99+ percent, Aldrich).

The metal stabilizer/$AlCl_3$ reactions were done similarly:

1. A measured amount of a PERC solution with a certain concentration of metal stabilizer and n-decane as an internal standard (99 g of 0.12 mole/L) were placed into a reaction vessel (a magnetically stirred Erlenmeyer flask).

2. The reaction vessel was placed into a constant-temperature water bath and the solution was allowed to reach the bath temperature.

3. An amount of $AlCl_3$ (anhydrous, Fisher Scientific) sufficient to react completely with the metal stabilizer (1 g) was weighed and added to the reaction vessel.

4. To monitor the concentration of metal stabilizer, aliquots were taken periodically from the vessel and filtered through filter paper using a Buchner funnel. The concentrations of the aliquots were determined by GC using the internal standard method.

The stabilizers used in these studies were 1,4-dioxane (high-performance liquid chromatography [HPLC] grade, 99.9 percent, Aldrich), 1,3-dioxolane (gold label, 99.5 percent, Aldrich), and nitromethane (gold label, 99+ percent, Aldrich).

For each kinetics study, the stoichiometry was found in the literature and then tested experimentally. An excess amount of inhibitor was reacted with HCl or $AlCl_3$ overnight and then the inhibitor concentration was measured using GC and the internal standard method. This concentration was used to determine the stoichiometry of the reactions.

Integral Method of Analysis of Batch Reaction Data. The batch reaction concentration data were used to determine the order and rate constant of the reaction. These reactions were treated as constant-volume batch reactions which were monitored by the disappearance of inhibitor.

In a constant-volume system, the rate expression for the disappearance of a compound, A, is defined by the following expression:

$$-r_A = -dC_A/dt = f(k,C) \qquad \text{[Eq 10]}$$

where: r_A = disappearance of A.

C_A = concentration of A.

t   = time.

k   = rate constant.

C   = concentration.

By assuming that the concentration-dependent terms can be separated from the concentration-independent terms, this expression results:

$$-dC_A/f(C) = k \, dt \qquad \text{[Eq 11]}$$

The function $f(C)$ can then be expressed in terms of $C=A$ and integrated to give:

$$-\int_{C_{AO}}^{C_A} dC_A/f(C_A) = \int_0^t k \, dt = kt \qquad \text{[Eq 12]}$$

The concentration function is proportional with time. Therefore, a plot of this function versus time results in a straight line of slope k. This method is extremely useful for fitting simple reaction types corresponding to elementary reactions.[48]

Temperature and Reaction Rate. For elementary reaction, Arrhenius' law predicts that the reaction rate constant varies with temperature as follows:

$$k \, \alpha \, e^{-E/RT} \qquad \text{[Eq 13]}$$

where: k = rate constant.
                E = activation energy.
                R = ideal gas law constant.
                T = absolute temperature.

Therefore, the rate constant can be expressed by the equation:

$$k = A \cdot e^{-E/RT} \qquad \text{[Eq 14]}$$

where A is the proportionality constant. By plotting the logarithms of the rate constant k ($\ln k$) values versus the inverse of the absolute temperature ($1/T$) the activation energy E and the Arrhenius constant A can be determined.[49]

Experimental Results

Identification of Solvent Inhibitors

TCE. Using GC-MS, five additives (impurities) were identified in the sample of new TCE received from Robins AFB. These results were further confirmed by spiking analysis when possible.

The first of these compounds was identified as butylene oxide (1,2-epoxybutane). This compound is an epoxide used as an acid acceptor in chlorinated solvents. It is an

---

[48]O. Levenspiel, *Chemical Reaction Engineering*, 2nd ed. (John Wiley and Sons, 1972), pp 41-86.
[49]O. Levenspiel.

extremely common additive in chlorinated solvents.[50]   Figure 37 compares the experimental and library mass spectra.   Results of GC-MS analysis were further confirmed by spiking.

The second of these compounds was identified as ethyl acetate (acetic acid, ethyl ester). It is used along with other chemicals to prevent the corrosion of metals by chlorinated solvents.[51]   Experimental and library mass spectra are compared in Figure 38. Again, these results were further confirmed by spiking.

A third compound was found to be 5,5-dimethyl-2-hexene. Figure 39 compares the experimental and library spectra.   No retention time analysis was performed because the pure compound was unavailable.   There was, however, excellent agreement between the mass spectra.

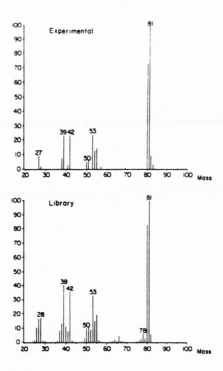

Figure 37.   Comparison of library and experimental mass spectra for identification of butylene oxide as a trichloroethylene additive/impurity.

[50]J. H. Rains, U.S. Patent 3,629,128 (1971); L. Peoples, U.S. Patent 3,746,648 (1973); N. L. Beckers and E. A. Rowe, U.S. Patent 3,935,287 (1976); N. L. Beckers and E. A. Rowe, U.S. Patent 3,957,893 (1976); L. S. McDonald, U.S. Patent 3,565,811 (1971); M. J. Culverland and H. R. Stopper, U.S. Publ. Patent Appl. B US 370,309 (1976); D. R. Spencer and W. L. Archer, U.S. Patent 4,115,461 (1978); C. L. Cormany, U.S. Patent 4,065,323 (1977).

[51]N. Ishibe and J. K. Harden, U.S. Patent 4,368,338 (1983).

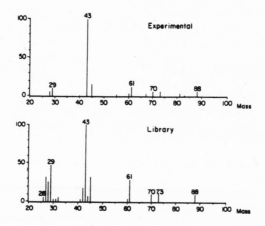

Figure 38.   Comparison of library and experimental mass spectra for identification of ethyl acetate as a trichloroethylene additive/impurity.

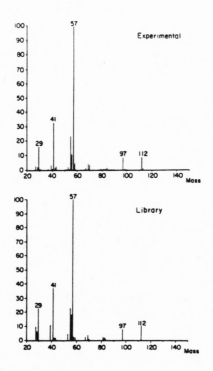

Figure 39.   Comparison of library and experimental mass spectra for identification of 5,5-dimethyl-2-hexene as a trichloroethylene additive/impurity.

The fourth compound was found to be epichlorohydrin (chloromethyloxirane), a known additive to chlorinated solvents. Like butylene oxide, it has an epoxide structure that suggests its action as an acid acceptor. It is a common additive in chlorinated solvents.[52]   Figure 40 compares the spectra. Retention time confirmation of these results was performed.

Finally, the fifth compound was found to be n-methylpyrrole. N-methylpyrrole, along with pyrrole and phenol, are often added as antioxidants to unsaturated chlorinated solvents (i.e., TCE and PERC).[53] The spectral comparison is shown in Figure 41. Again, spiking analysis was performed.

The inhibitor package for TCE is composed of acid acceptors and antioxidants. Table 15 summarizes the inhibitors identified and Figure 42 is a labeled chromatogram.

PERC. The GC-MS analysis identified two additives in the sample of new PERC supplied by Kelly AFB. The first of these was cyclohexene oxide (7-oxabicyclo-[4.1.0]heptane). It has an epoxide-type structure that suggests it is an acid acceptor. It

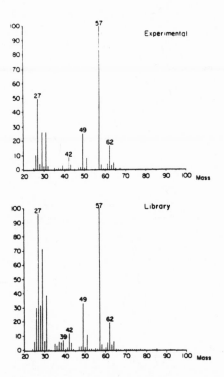

Figure 40.   Comparison of library and experimental mass spectra for identification of epichlorohydrin as a trichloroethylene additive/impurity.

[52]J. A. Manner, U.S. Patent 3,532,761 (1970); W. L. Archer, E. L. Simpson, and R. R. Gerard, U.S. Patent 4,018,837 (1977).
[53]J. W. Tipping, Brit. GB 1,276,783 (1972); J. A. Borror and E. A. Rowe, Jr., U.S. Patent 4,293,433 (1981).

is a known additive to PERC.[54] The spectrum for cyclohexene oxide was not in the MS spectra library, and therefore, this compound was initially identified by visual comparison to another spectrum library.[55] This identification was then confirmed by comparing the mass spectrum for a known sample of cyclohexene oxide to the spectrum attained in the PERC sample analysis. The result was further confirmed using retention time analysis. The comparison of mass spectra is shown in Figure 43.

The second compound was identified as butoxymethyl oxirane. Like cyclohexene oxide, it has the epoxide structure of an acid acceptor. Figure 44 compares spectra. Confirmation by spiking was not performed due to unavailability of the pure compound. There was, however, extremely good agreement between the library and experimental mass spectra.

An interesting finding in this identification was the absence of an antioxidant. This finding may be due to the very stable nature of PERC. A labeled chromatogram is shown in Figure 45. Table 16 summarizes the additives identified.

MC. Three additives were identified by the GC-MS analysis of new MC supplied by Robins AFB. The first of these was n-methoxymethanamine, an amine compound that probably acts as an acid acceptor. Figure 46 compares the experimental and library mass spectra. Retention time analysis was not performed because pure compound was unavailable, but the mass spectral agreement was excellent.

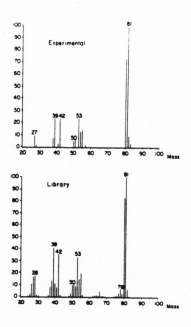

Figure 41.   Comparison of library and experimental mass spectra for identification of n-methylpyrrole as a trichloroethylene additive/impurity.

[54]B. Van Gemert, Brit. UK Pat. Appl. GB 2,027,697 (1980).
[55]E. Stenhagen, et al., *Registry of Mass Spectral Data* (John Wiley and Sons, 1974).

Table 15

Additives/Impurities Identified in Trichloroethylene

Inhibitor	Formula (MW)	Structure	BP (°C)	Function
Butylene oxide	C_4H_8O (72.1)	$CH_3CH_2\overset{O}{\overset{\mid}{CH}}-CH_2$	63.3	Acid acceptor
Ethyl acetate	$C_4H_8O_2$ (88.1)	$CH_3\overset{O}{\overset{\mid}{C}}OCH_2CH_3$	77.06	Unknown
5,5 Dimethyl-2-hexene	C_8H_{16} (112.2)	$CH_3\overset{CH_3}{\underset{CH_3}{\overset{\mid}{\underset{\mid}{C}}}}CH_2CH_2CH=CH_2$	Unknown	Unknown, possibly antioxidant
Epichloro-hydrin	C_3H_5OCl (92.5)	$ClCH_2\overset{O}{\overset{\diagup\diagdown}{CH-CH_2}}$	16.5	Acid acceptor
n-Methyl-pyrrole	C_5H_7N (81.1)	(ring with N and CH_3)	114.5	Antioxidant

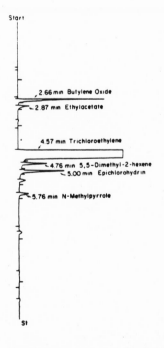

Figure 42. Gas chromatogram of trichloroethylene.

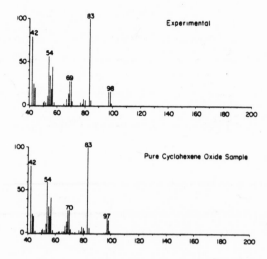

Figure 43.   Comparison of the experimental mass spectra with those of cyclohexene oxide for identification of cyclohexene oxide as a tetrachloroethylene additive.

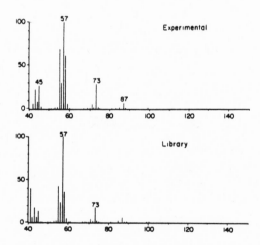

Figure 44.   Comparison of library and experimental mass spectra for identification
of butoxymethyl oxirane as a tetrachloroethylene additive.

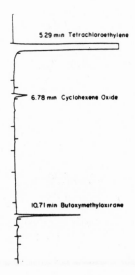

5.29 min Tetrachloroethylene

6.78 min Cyclohexene Oxide

10.71 min Butoxymethyloxirone

Figure 45. Gas chromatogram of tetrachloroethylene.

Table 16

Additives Identified in Tetrachloroethylene

Inhibitor	Formula (MW)	Structure	BP (°C)	Function
Cyclo-hexene oxide	$C_6H_{10}O$ (98.2)		131.5	Acid Acceptor
Butoxy-methyl oxirane	$C_7H_{14}O_2$ (130.2)	$CH_3(CH_2)_3OCH_2CH\text{-}CH_2$	Unknown	Acid Acceptor

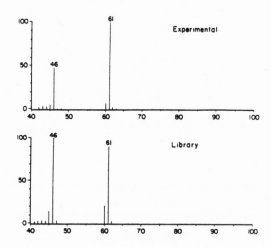

Figure 46.   Comparison of library and experimental mass spectra for identification of n-methoxymethanamine as a 1,1,1-trichloroethane additive.

A second additive identified was formaldehyde dimethyl hydrazone. Hydrazones are known additives in chlorinated solvents.[56]   Their action is probably as Al/AlCl$_3$ complexing agents. The spectral comparison is shown in Figure 47. The pure compound was unavailable and therefore no spiking was performed. There is, however, good agreement between the mass spectra.

The final inhibitor identified in the MC sample was 1,4-dioxane. This inhibitor is an extremely common additive in MC.[57] Its primary function is as a metal stabilizer. For this sample, 1,4-dioxane was identified by visual comparison to a library not contained within the MS system.[58] The identification was then verified by performing MS analysis of a known 1,4-dioxane sample and comparing this spectrum with the one obtained from MS analysis of the MC. Verification was performed using spiking analysis. The spectral comparison can be seen in Figure 48.

The identified inhibitor package is very close to that expected for MC. The additives are mainly metal stabilizers with some acid acceptors. Figure 49 is a labeled chromatogram for the MC sample. Table 17 summarizes the inhibitors.

[56]N. L. Beckers, U.S. Patent 3,796,755 (1974).
[57]J. H. Rains; D. R. Spencer and W. L. Archer; H. Richtzenhain and R. Stephan, U.S. Patent 3,787,509 (1974); H. Richtzenhain and R. Stephan, U.S. Patent 3,959,397 (1976); J. A. Manner, U.S. Patent 4,026,956 (1977).
[58]O. Levenspiel.

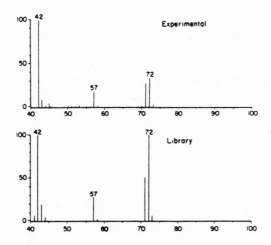

Figure 47. Comparison of library and experimental mass spectra for identification of formaldehyde dimethyl hydrazone as a 1,1,1-trichlorethane additive.

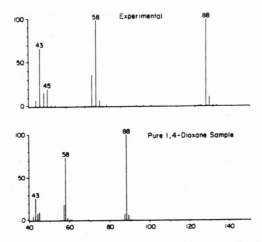

Figure 48. Comparison of 1,4-dioxane mass spectrum with experimental mass spectrum for identification of 1,4-dioxane as a 1,1,1-trichloroethane additive.

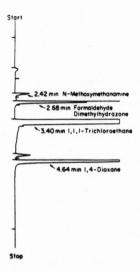

Figure 49. Gas chromatogram of 1,1,1-trichloroethane.

Table 17

Additives Identified in 1,1,1-Trichloroethane

Inhibitor	Formula (MW)	Structure	BP (°C)	Function
N-Methoxy-methanamine	C_2H_7NO (61.1)	$CH_3OCH_2NH_2$	Unknown	Acid acceptor
Formaldehyde dimethyl hydrazone	$C_3H_8N_2$ (72.1)	$CH_2=NN\begin{smallmatrix}CH_3\\CH_3\end{smallmatrix}$	Unknown	Al stabilizer
1,4-Dioxane	$C_4H_8O_2$ (88.1)		101	Al stabilizer

Solvent Inhibitor Time Studies

Inhibitor concentration analysis (GC with internal standard) was performed on solvent samples (TCE, PERC, and MC) taken at regular intervals during the vapor degreasing operation cycle. The concentration data were then plotted versus time (concentration as the ordinate, time as the abscissa). These graphs were used to determine the behavior of solvent inhibitors over time in a typical vapor degreasing operation. The data were also used to determine the inhibitor level in the spent solvent (i.e., that for disposal or recycling).

TCE. Samples for the time study of TCE solvent inhibitor concentration were provided by Robins AFB. The samples were taken periodically during the normal operational cycle of a typical vapor degreaser from the time new solvent was added until it was spent.

The graph of inhibitor concentration versus time, Figure 50, shows that there was considerable depletion of the acid acceptance inhibitor butylene oxide (about 50 percent). Also, there were slight changes in concentrations of the other additives (ethyl acetate, epichlorohydrin, and N-methylpyrrole). In addition, the concentration-versus-time curve for butylene oxide and the acid acceptance-versus-time curve for TCE (Figure 14) are very similar. Finally, the AAV of the spent solvent was considerably larger than the minimum recommended value (0.085 weight percent NaOH compared with 0.04 weight percent NaOH).

PERC. Samples of PERC were taken periodically during two cleaning cycles (PERC#1 and PERC#2) of a typical vapor degreasing operation at Kelly AFB. These samples ranged from new to spent solvent. Inspection of the concentration-versus-time curves for PERC (Figures 51 and 52) shows a large increase in butoxypropylene oxide concentrations (about 130 percent in each cycle). The cyclohexene concentration in PERC#1 was unchanged. There was, however, a change in the concentration of cyclohexene oxide in PERC#2 (about a 22 percent decrease).

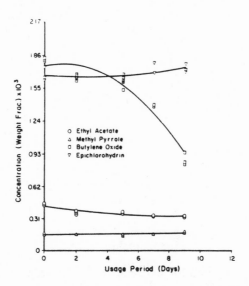

Figure 50. Variation in inhibitor concentration of trichloroethylene with usage time.

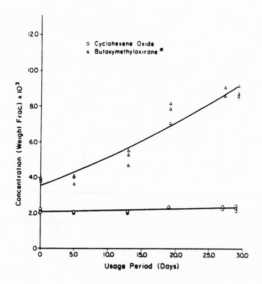

*Due to lack of the pure compound, the concentration of butoxymethyl oxirane was calculated assuming a response factor of 1.

Figure 51.   Variation in inhibitor concentration of tetrachloroethylene with usage time (PERC#1).

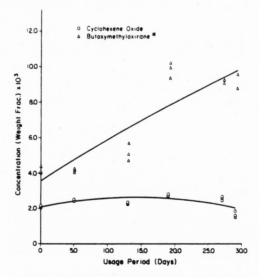

*Due to lack of the pure compound, the concentration of butoxymethyl oxirane was calculated assuming a response factor of 1.

Figure 52.   Variation in inhibitor concentration of tetrachloroethylene with usage time (PERC#2).

These results compare favorably with those for the acid acceptance studies for PERC (Figure 23). An interesting finding was the considerably larger AAV for the spent solvent samples compared with new solvent (0.19 vs. 0.12 weight percent NaOH in PERC#1; 0.16 vs. 0.13 percent NaOH in PERC#2). This result was probably due to the fact that the additives (cyclohexene oxide and butoxymethyl oxirane) are less volatile than PERC and thus would become concentrated as the solvent evaporated.

To test this theory, a simple experiment was performed. A sample of inhibited PERC was partially distilled and the distillate and bottoms were analyzed for cyclohexene oxide and butoxymethyl oxirane concentration. The distillate was found to have a considerably lower concentration of the two additives compared with the original concentration, whereas the bottoms showed a substantially higher concentration of inhibitors; thus, it was shown that the additives were less volatile than PERC.

MC. Samples of MC were taken periodically during the cleaning cycle (new to spent solvent) of a typical vapor degreasing operation at Robins AFB. Makeup solvent was added after sample 4.

Figure 53 shows a significant change in 1,4-dioxane concentration in the period between samples 1 and 4 (a 37 percent decrease). The 1,4-dioxane concentration approached that of new solvent after the addition of fresh solvent and remains basically constant until the end of the cleaning cycle. Figure 54 shows that formaldehyde dimethyl hydrazone concentration remained fairly constant during the cycle. Figure 54 shows a gradual rise (44 percent) in n-methoxymethanamine in samples 1 through 4 with a return to original levels after solvent addition. There was also a gradual rise in n-methoxymethanamine concentration during the rest of the cycle (25 percent). (The variation in AAV for MC with usage time was shown in Figure 30.)

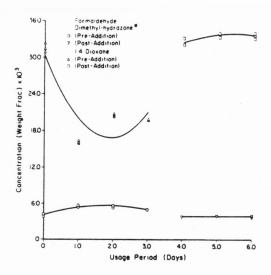

*Due to lack of pure compound, the formaldehyde dimethyl hydrazone concentration was calculated assuming a response factor of 1.

Figure 53.   Variation in inhibitor concentration of 1,1,1-trichloroethane with usage time.

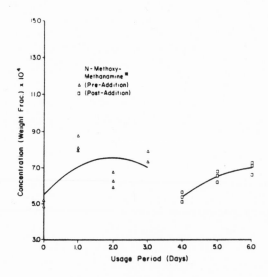

*Due to lack of pure compound, the n-methoxymethanamine concentration was calculated assuming a response factor of 1.

Figure 55.   Variation in inhibitor concentration of 1,1,1-trichloroethane with usage time.

Reclaimed Solvent Inhibitor Studies

The purpose of these studies was to determine the effects of common reclamation methods on the inhibitor concentration of the solvents (TCE, PERC, and MC). Used solvent samples were treated using distillation or adsorption over activated carbon. These "reclaimed" solvent samples were then analyzed using GC and the internal standard method. The inhibitor concentrations of the reclaimed solvent were compared with those of new and used solvents.

TCE. Spent TCE samples obtained from Robins AFB were treated using distillation or activated carbon adsorption. The inhibitor concentrations of reclaimed solvents were then determined and compared with those of new and spent TCE from the same source. The inhibitor concentrations of new, spent, and reclaimed solvents are given in Table 18.

Comparison of the inhibitor concentrations of distilled and spent TCE shows that distillation was effective in recovering the inhibitors present in TCE since butylene oxide, ethyl acetate, and epichlorohydrin were readily reclaimed with the TCE distillate. There was, however, some decrease in methyl pyrrole concentration resulting from distillation, probably due to its considerably higher boiling point than TCE. Nevertheless, the distilled solvent's methyl pyrrole concentration was similar to that of new solvent. It also was noted that the epichlorohydrin level of the distilled solvent was approximately the same as that of new TCE, whereas the concentrations of butylene oxide and ethyl acetate were considerably lower than those of new solvent. However, comparison of the inhibitor levels of solvent treated by activated carbon adsorption and spent TCE shows a selective adsorption of butylene oxide, epichlorohydrin, and methyl pyrrole. There was, though, little difference between the ethyl acetate levels of carbon-adsorbed and spent TCE. Moreover, the comparison of reclaimed solvent to new TCE indicates that the concentrations of butylene oxide, ethyl acetate, epichlorohydrin, and methyl pyrrole were considerably lower than their respective concentrations in new solvent.

PERC. Spent PERC samples obtained from Kelly AFB were treated by distillation and adsorption with activated carbon. The inhibitor concentrations of these reclaimed solvents were then determined and compared with those of new and used tetrachloroethylene from the same source. The inhibitor concentrations of new, used, and reclaimed solvents are listed in Table 19.

The comparison of inhibitor concentrations between used and distilled PERC shows that cyclohexene oxide was readily recovered with the PERC distillate. The concentration, however, of butoxymethyl oxirane in the reclaimed solvent was considerably lower than that in spent PERC, probably due to the lower volatility of butoxymethyl oxirane relative to tetrachloroethylene. Nevertheless, the concentration of butoxymethyl oxirane in the distilled solvent was higher than its concentration in new PERC. Also, the cyclohexene oxide level in the distilled solvent was approximately the same as that in new PERC.

Table 18

Inhibitor Concentrations of New, Spent, and Reclaimed Trichloroethylene*

Sample	Inhibitor Concentration (Weight Fraction)			
	Butylene Oxide $(\times 10^3)$	Epichloro-hydrin $(\times 10^3)$	Ethyl Acetate $(\times 10^4)$	Methyl Pyrrole $(\times 10^4)$
New TCE	1.64	1.66	3.46	1.59
Spent TCE	0.685	1.69	2.85	2.18
TCE (run #1) distillate	0.717	1.55	2.50	1.65
TCE (run #2) distillate	0.719	1.67	2.66	1.68
Carbon-adsorbed TCE (run #1)	0.352	1.25	2.59	0.833
Carbon-adsorbed TCE (run #2)	0.528	1.37	2.71	0.966

*The internal standard used was 2×10^{-3} weight fraction tetrahydrofuran.

Table 19

Inhibitor Concentrations of New, Spent, and Reclaimed Tetrachloroethylene*

Sample	Inhibitor Concentration (Weight Fraction)	
	Cyclohexene Oxide ($\times 10^3$)	Butoxymethyl Oxirane** ($\times 10^3$)
New PERC	1.06	4.26
Spent PERC	0.988	7.45
PERC distillate (run #1)	0.972	5.32
PERC distillate (run #2)	0.961	5.52
Carbon-adsorbed PERC (run #1)	0.182	5.59
Carbon-adsorbed PERC (run #2)	0***	5.21

*The internal standard used was 2×10^{-3} weight fraction 1,4-dioxane.
**Due to the unavailability of pure compound, the butoxymethyl oxirane concentration was estimated using GC and the internal standard method assuming a response factor of unity, thus allowing a comparison of its concentration in each sample.
***Compound concentration below detectable limit.

PERC reclaimed by adsorption with activated carbon showed an extreme decrease in cyclohexene oxide level compared with used solvent. In addition, it saw a considerable decrease in butoxymethyl oxirane concentration in comparison to the butoxymethyl oxirane concentration of spent PERC. Also, comparison of carbon-adsorbed solvent with new solvent shows that the cyclohexene oxide level was much lower in the reclaimed solvent than in new PERC. The butoxymethyl oxirane concentration of the carbon-adsorbed solvent was greater than that in new PERC.

MC. Spent MC samples obtained from Hayes International Corp. were reclaimed using distillation and activated carbon adsorption. The inhibitor concentrations in these reclaimed MC samples were determined and then compared with those of new and spent MC obtained from the same source. The inhibitor concentrations of new, used, and reclaimed MC can be seen in Table 20.

The comparison of inhibitor concentrations between distilled and spent MC demonstrates that n-methoxymethanamine and formaldehyde dimethyl hydrazone were readily recovered with the MC distillate. There was some loss of 1,4-dioxane by distillation, probably due to its higher boiling point than MC. The comparison of 1,4-dioxane concentration in distilled MC to that of new solvent, however, shows that the formaldehyde dimethyl hydrazone levels of distilled solvent were greater than those of new MC, whereas the level of n-methoxymethanamine in the reclaimed solvent was considerably lower than that in new MC.

Comparison of spent MC to carbon-adsorbed solvent shows that the treated solvent had considerably lower concentrations of methoxymethanamine, formaldehyde dimethyl hydrazone, and 1,4-dioxane. Also, when solvent reclaimed by activated carbon adsorption is compared with new MC, the reclaimed solvent's n-methoxymethanamine and formaldehyde dimethyl hydrazone levels are seen to be significantly lower than those of new solvents; in contrast, the 1,4-dioxane level of carbon-adsorbed solvent was greater than that found in new MC.

Inhibitor Kinetic Studies

Batch reactor kinetic studies were performed to obtain more familiarity with the characteristics of additives common to chlorinated solvents. The data from these experiments were used to determine the order, rate constant, activation energy, and Arrhenius constant of the reactions. The reactions studied were of two general types: acid acceptor/HCl and metal stabilizer/$AlCl_3$.

Table 20

Inhibitor Concentrations of New, Spent, and Reclaimed 1,1,1-Trichloroethane*

Sample	Inhibitor Concentration (Weight Fraction)		
	n-Methoxy-1 methanamine $(x\ 10^4)$**	Formaldehyde Dimethyl Hydrazone $(x\ 10^3)$**	1,4-Dioxane $(x\ 10^3)$
New MC	8.92	5.78	17.2
Spent MC	4.14	6.16	29.0
MC distillate (run #1)	4.48	7.05	20.1
MC distillate (run #2)	4.71	7.42	19.1
Carbon-adsorbed MC (run #1)	1.54	3.35	23.9
Carbon-adsorbed MC (run #2)	1.07	3.39	22.9

*The internal standard used was 5 x 10^{-3} weight fraction tetrahydrofuran.
**Due to the unavailability of pure compounds, the N-methoxymethanamine and formaldehyde dimethyl hydrazone concentrations were determined using GC and the internal standard method assuming a response factor of unity, thus allowing comparison of their concentrations in each sample.

Acid Acceptor/HCl Reactions. These experiments were done using three common epoxide acid acceptors: butylene oxide, epichlorohydrin, and cyclohexene oxide. PERC was used as a solvent due to its stability. The PERC/acid acceptor mixture was placed into a reaction vessel (a glass vial with a septum cap) and kept at a constant temperature, Tc, using a shaker bath. At a time designated as "time zero," a stoichiometric quantity (determined from literature and experimentally) of a nonaqueous HCl (HCl in isopropanol) at Tc was added as a slug to the reaction vessel. The concentration of the acid acceptor was then monitored over time using GC and the internal standard method. This time and concentration data were then used to determine the order and rate constant of the reaction. Batch reaction experiments of each acid acceptor were performed at three distinct temperatures to determine the temperature dependence of their reaction with HCl.

In the integral analysis of the batch reaction data, it was assumed initially that the acid acceptor/HCl reactions were bimolecular and second order overall. This assumption was based on the one-to-one stoichiometry (HCl to acceptor) of the reactions (butylene oxide, epichlorohydrin, and cyclohexene oxide with HCl).

Once it is assumed that a reaction is second-order bimolecular with a stoichiometry of 1:1 or:

$$A + B \longrightarrow P \qquad \qquad \text{[Eq 15]}$$

the depletion rate of A can be expressed in terms of concentration as:

$$-r_A = -dC_A/dt = k\, C_A C_B \qquad \qquad \text{[Eq 16]}$$

where: C_i = the concentration of i (A, B).
           t   = time.
           k   = reaction rate constant.

If $C_A = C_B$, then this equation can be written as:

$$-dC_A/dt = k\, C_A^2 \qquad \qquad \text{[Eq 17]}$$

Integrating Equation 17 produces:

$$1/C_A = kt + 1/C_{A0} \qquad \qquad \text{[Eq 18]}$$

Therefore, a plot of $1/C_A$ versus time will result in a straight line of slope k and an intercept of $1/C_{A0}$.

The concentration-versus-time data for the respective batch reaction between the acid acceptors and HCl, when plotted as if they were second—order, approximated a straight line. The plots of $1/C_A$ versus time for the reactions of HCl with butylene oxide, epichlorohydrin, and cyclohexene oxide, respectively, are shown in Figures 55 through 57.

The Arrhenius equation:

$$k = Ae^{-E/RT}$$

can be used to plot the logarithm (ln) of k versus 1/T to obtain a straight line of slope -E/R and an intercept of ln A. Figures 58 through 60 show the Arrhenius plots for the respective reactions of butylene oxide, epichlorohydrin, and cyclohexene oxide with HCl.

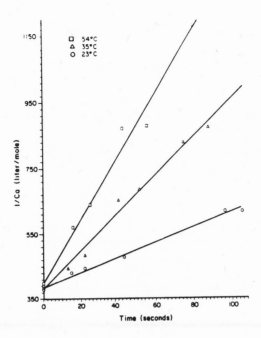

Figure 55. Integral analysis test for a second-order reaction: butylene oxide reacting with HCl.

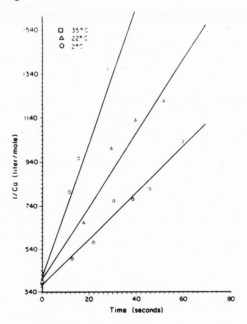

Figure 56. Integral analysis test for second-order reaction: cyclohexene oxide reacting with HCl.

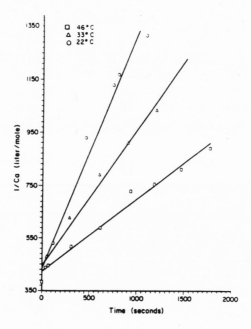

Figure 57. Integral analysis test for a second-order reaction: epichlorohydrin reacting with HCl.

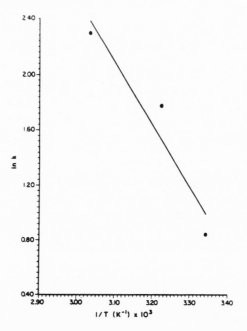

Figure 58. Determination of the temperature dependence of the butylene oxide/ HCl reaction using Arrhenius' law.

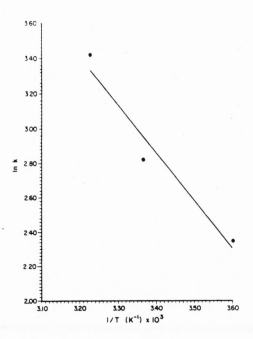

Figure 59. Determination of the temperature dependence of the cyclohexene oxide/HCl reaction using Arrhenius' law.

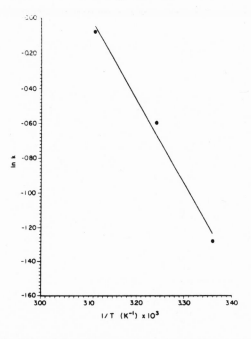

Figure 60. Determination of the temperature dependence of the epichlorohydrin/HCl reaction using Arrhenius' law.

Metal Stabilizer/AlCl$_3$ Reactions. The stabilizers used were 1,4-dioxane, 1,3-dioxolane, and nitromethane. The solution of stabilizer and PERC was placed in a reaction vessel and brought to constant temperature, Tc, in a water bath. The reaction vessel used was a magnetically stirred Erlenmeyer flask. A stoichiometric amount of AlCl$_3$ (determined from the literature and experimentally) was added at time zero. The reaction vessel was kept well mixed to provide a uniform distribution of the suspended AlCl$_3$. Aliquots were taken from the reaction vessel and filtered through filter paper using a Buchner funnel. The concentrations were measured by GC and the internal standard method. Each reaction was performed at three distinct temperatures.

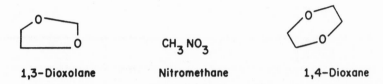

| 1,3-Dioxolane | Nitromethane | 1,4-Dioxane |

The stoichiometry of each AlCl$_3$ complexation reaction was found to be 1:1. Therefore, it was assumed that the reactions were bimolecular and second order. The data for each stabilizer/AlCl$_3$ reaction were then plotted as $1/C_A$ versus time. For dioxane and dioxolane, these plots approximated a straight line (Figures 61 and 62).

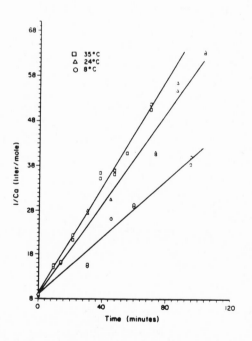

Figure 61.   Integral analysis test for a second-order reaction: 1,4-dioxane reacting with AlCl$_3$.

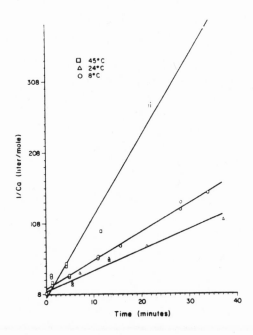

Figure 62. Integral analysis test for a second-order reaction: 1,3-dioxolane reacting with AlCl$_3$.

The metal stabilizer/AlCl$_3$ reactions were performed similarly.   For the nitromethane/AlCl$_3$ reactions, the second-order plot was not a straight line (Figure 63).   This reaction was difficult to monitor due to a fast reaction rate (90 percent reacted during the first minute).   This fast rate, coupled with the need to filter the reaction solids, made it nearly impossible to obtain accurate data.

The rate constant values obtained from analyzing the batch reactor data can then be plotted versus temperature in an Arrhenius plot to determine the activation energy and Arrhenius constant for the respective stabilizer/AlCl$_3$ reactions.   These plots are shown in Figures 64 and 65.

Metal Stabilizer Test

The different types of inhibitors are lost from chlorinated solvents to a varying degree, depending on the type of operation.   For example, if HCl is produced in unusually high concentrations, the acid acceptors and the antioxidants may be depleted from the solvent.   If aluminum fines from the work (parts) or degreaser equipment metal are present in the degreaser at a high enough level, then the metal stabilizers are consumed rapidly from the solvent.   Thus, quantitative tests to determine the concentration of the different types of inhibitors are necessary in monitoring the quality of a solvent.   A quantitative test for acid acceptors is the AAV test (ASTM D 2942).   This test was been described earlier in this chapter.   No simple test was found in the literature for the metal stabilizers specifically, except for an "aluminum scratch test" for MC (ASTM D 2943).   The metal stabilizers detected by GC/MS in the solvents used by DOD installations were ethylacetate, formaldehyde dimethyl hydrazone, and 1,4-dioxane.   A brief description of the quantitative tests determined from the literature for these metal stabilizers follows.

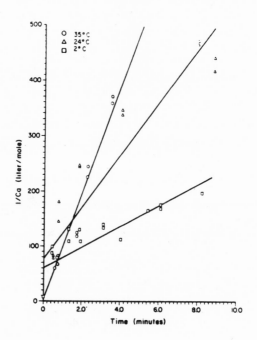

Figure 63.   Integral analysis test for a second-order reaction: nitromethane react-ing with AlCl₃.

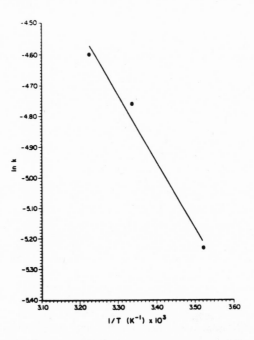

Figure 64.   Determination of the temperature dependence of the 1,4-dioxane/AlCl₃ reaction using Arrhenius' law.

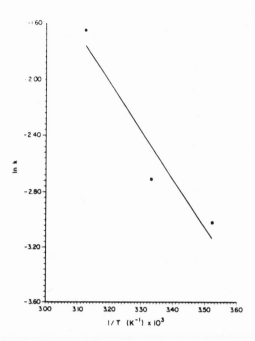

Figure 65.   Determination of the temperature dependence of the 1,3–dioxolane/AlCl$_3$ reaction using Arrhenius' law.

Ethyl Acetate. A common way of determining ethyl acetate concentration is by titration. The reagents used for titration are aluminum ethyl and phenyl and their chloro and bromo derivatives, e.g., $(C_2H_5)_2AlCl$, and $(C_2H_5)AlCl_2$. A nonaqueous solvent such as toluene, xylene, or heptane is used to dissolve the aluminum compound. The endpoint determination may be potentiometric or via color change of a triphenylmethane dye such as methyl, crystal, or gentian violet.[59] The product is a complex formed by semipolar bond formation between an acetate oxygen atom and aluminum receptor atom. This complex formation suggests that ethyl acetate is added to chlorinated solvents for its metal-stabilizing characteristic.

Other reactions specific to acetates (esters) are hydrolysis, transesterification, ammonolysis, Grignard reagent reaction, and reduction reactions, either by catalytic hydrogenation or chemical reduction.[60]

Formaldehyde Dimethyl Hydrazone. This compound has been known to be used as a stabilizer in chlorinated hydrocarbons.[61] A stabilizer must be only a mild base so as to preclude attack on metals. Thus, both the acidity from solvent decomposition and the basicity from high concentrations of stabilizers are equally detrimental.

---

[59]M. R. F. Ashworth, *Titrimetric Organic Analysis, Part I: Direct Methods* (Interscience Publishers, 1964).

[60]C. D. Gutsche and D. J. Pasto; A. S. Wingrove and R. L. Caret, *Organic Chemistry* (Harper and Row, 1981).

[61]K. Johnson.

The degreasing of metal is generally done in the vapor phase and adequate stabilizer has to be present in both the vapor and liquid phases. This condition is most important for trichloroethylene vapor which is highly susceptible to breakdown by atmospheric oxygen.[62] All of the above characteristics for a stabilizer are satisfied by a formaldehyde dialkyl hydrazone, e.g., formaldehyde dimethyl hydrazone:

$$CH_2 = N - N \begin{array}{c} CH_3 \\ CH_3 \end{array}$$

FORMALDEHYDE DIMETHYL HYDRAZONE

The literature reports tests to assess the effect of formaldehyde dimethyl hydrazone on solvent performance. A solvent containing 0.025 percent by weight of formaldehyde dimethyl hydrazone was subjected to an accelerated oxidation test, and then the solvent was extracted with an equal volume of water. The water extract pH was found to be 7.0. The solvent accelerated oxidation test procedure was published in Army-Navy Aeronautical Specification U.S. MIL-T-7003 of 5 September 1950.[63] In accordance with this standard, 200 mL of a stabilized solvent were subjected to 48 hr of reflux boiling into a stream of water-saturated oxygen using a 150-W lamp as the heat source. Steel plates were placed into the liquid and vapor phases to determine the corrosivity, and the test liquid was then extracted with water to test the pH. The phosgene content was determined from the test liquid by conventional analysis. When the formaldehyde dimethyl hydrazone content was reduced to 0.01 percent by weight, only a slight reduction in water extract pH was determined after an accelerated oxidation test.

A titration scheme has been formulated for detecting the concentration of hydrazine substitution products--a class to which formaldehyde dimethyl hydrazone belongs. The reagent used is ethanol and the endpoint is potentiometric.[64]

1,4-Dioxane. Titrations done with: tertiary bases (pyridine and related compounds, dimethylaniline, and diethylamine); ethers such as dioxane; ketones such as acetone and benzophenone; ethyl acetate; and metal ethoxides against halides in nonaqueous solvents result in complex formation, or neutralization product of Lewis acids and bases.[65] The halide reactants are chlorides and bromides of elements from periodic groups 2 through 8 (e.g., $AlCl_3$) in solvents such as heptane, benzene, and carbon tetrachloride. The endpoint determination is via a color change of indicators such as crystal violet, malachite green, and benzanthrone.[66]

Many authors have studied the effect of 1,4-dioxane together with other metal stabilizers and acid acceptors. Some experiments consisted of refluxing inhibitor containing solvent in a round-bottomed flask equipped with a Soxhlet extractor and topped with a reflux condenser. An aluminum strip placed in a thimble in the extractor was always in contact with the condensed solvent. The strip was weighed after 3 days and the extent of lost aluminum correlated with inhibitor concentration.[67]

---

[62]K. Johnson.
[63]K. Johnson.
[64]M. R. F. Ashworth.
[65]M. R. F. Ashworth.
[66]M. R. F. Ashworth.
[67]W. L. Archer and E. L. Simpson; W. L. Archer.

Another test found in the literature requires that constant heat be supplied to an inhibited solvent and maintained at approximately 80 °C. A thermometer is suspended in the liquid to monitor heat requirements. A cleaned piece of aluminum wire and an aluminum metal strip (2.5 x 18 in.) are kept immersed in the solvent, and corrosion effects are correlated with time of contact.[68]

Another well known metal stabilizer is 1,3-dioxolane, the cyclic equivalent of ethylene gylcol.[69] This compound has the formula:

$$H_2C \underline{\quad\quad} CH_2$$

O   O

CH$_2$

1,3-DIOXOLANE

The amount of 1,3-dioxolane used is generally within 1 to 3 percent by weight solvent. Addition of phenolic antioxidants (p-t-butyl phenol; 2,6-di-t-butyl-p-cresol; or nonylphenol) in a very small quantity (0.01 to 1 percent) with 1,3-dioxolane yields improved decomposition protection to chlorinated solvents. In the absence of 1,3-dioxolane, the phenolic antioxidants have no effect at all.[70]

The comparison between 1,4-dioxane and 1,3-dioxolane as decomposition inhibitors indicates that 1,3-dioxolane is a superior inhibitor. Thus, relative to 1,4-dioxane, a lower concentration of 1,3-dioxolane would be required to afford equivalent decomposition protection to a solvent. The tests conducted to gauge the performance of 1,3-dioxolane are similar to those for 1,4-dioxane.[71]

Experimental. Two different methods were used in an attempt to formulate a test for metal stabilizer. The procedures for these two methods are described below.

Method 1. A measured amount of AlCl$_3$ was added to a known volume of solvent. HCl evolved as a decomposition product (Figure 67), with the amount generated a function of the amount of metal stabilizer in the solvent. The reaction was run in a Paar oxygen bomb calorimeter that was sealed as soon as all reactants had been added. The evolved HCl was measured on a pressure gauge fitted on the calorimeter. Figure 66 illustrates the results of three experimental runs using the following reaction parameters:

Volume of container + fittings	350 mL
Temperature	30 °C
AlCl$_3$	1.03 g
Solvent (1,1,1 trichlorethane)	25 mL (≈33 g)
Reaction time	3 hr

The plot in Figure 66 shows the pressure exerted by HCl vapor as a function of 1,4-dioxane present in the solvent. It indicates that HCl generation decreases significantly with increasing amounts of 1,4-dioxane.

---

[68]K. Johnson.
[69]K. Johnson; W. L. Archer; W. L. Archer and E. L. Simpson.
[70]K. Johnson.
[71]K. Johnson.

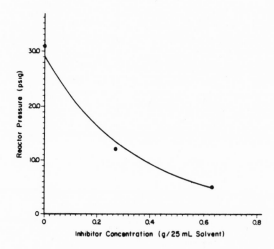

Figure 66. Effect of 1,4-dioxane on reactor pressure.

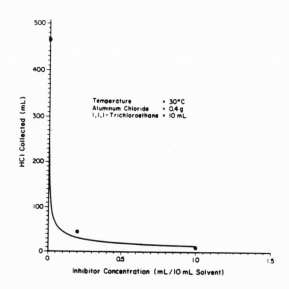

Figure 67. Effect of 1,4-dioxane on HCl collected.

Method 2. This method is essentially the same as Method 1, except for the mechanism of HCl detection. In Method 2, the HCl generated was collected in an inverted buret, initially filled with paraffin oil. The HCl gas was collected in the buret by displacing the paraffin oil. The results are shown in Figure 67. Here also, the effect of 1,4-dioxane on the generation of HCl was significant. More experimental work is required on these methods to fine-tune a test that would reliably quantify the metal stabilizer content in a chlorinated solvent. Further research is required for (1) detection of small changes in metal stabilizer concentration (i.e., increasing sensitivity of the methods), (2) observing individual effects of different stabilizers followed by a study of synergistic effects exhibited when multiple inhibitors are added, and (3) studying the behavior of commercial products (e.g., DOD solvents and products obtained from Dow Chemical Company and Aldrich Chemical Company).

Discussion

At present, installations that do vapor degreasing produce a great deal of hazardous wastes. With the high cost of new solvent as well as disposal of waste, it is extremely important to monitor the quality of these solvents and thus minimize the production of wastes. Current criteria for disposal or reclamation of these solvents are rather arbitrary and lead to unnecessary disposal or reclamation costs.

A more scientific approach to monitoring solvent quality could lead to substantial savings. Moreover, steps are now being taken to regulate the evaporative loss of these cleaning solvents. Present disposal costs are minimized to some extent due to solvent loss. However, effective prevention of solvent emissions would greatly increase the need to minimize waste production and would make solvent reclamation an extremely attractive option, thus increasing the importance of monitoring solvent quality. The tests described in this chapter have led to several useful conclusions with respect to assessing solvent quality. These findings are summarized below.

Identification of Solvent Inhibitors

Perhaps the main parameter in determining solvent quality is the inhibitor concentration. Inhibitors can be classified into three basic groups: acid acceptors, antioxidants, and metal stabilizers. The acid acceptors are usually epoxide or amine compounds that neutralize any HCl formed due to the breakdown of these chlorinated solvents. The antioxidants are generally aromatic compounds with a phenol and/or an amine side group which prevent autoxidation of unsaturated chlorinated solvents (TCE and PERC). Finally, the metal stabilizers prevent the violent degradation of these solvents when they contact aluminum or other metals (e.g., zinc and magnesium). These compounds perform two major functions: (1) they compete with the solvent for free metal sites and perform an insoluble protective deposit on the metal surface and (2) they react with any free metal chlorides that may be formed by solvent/metal reactions. The inhibitor package is determined by the solvent's physical characteristics, the nature of the parts to be cleaned, and the demands of the cleaning process. In general, MC contains mainly metal stabilizers and acid acceptors whereas TCE and PERC additives are mainly acid acceptors and antioxidants.

It is difficult to identify the acid acceptors in industrial chlorinated solvents because they are proprietary. For this study, it was necessary to perform this analysis using GC-MS. The compounds were identified successfully in this way (Tables 15 through 17). Identification of these compounds allowed USA-CERL to monitor the inhibitor concentration with time, study the effect of common reclamation techniques on the solvent concentration, and determine the kinetic parameters of some solvent additives.

Solvent Inhibitor Time Studies

The inhibitor concentration-versus-time studies (Figures 50 through 54) demonstrated the complex nature of the relationship between inhibitor level and usage time. Changes in the additive levels may have been due to inhibitor reactions or the difference in the volatilities of the solvent and additives. The major conclusion that can be made from these studies is that each of the spent solvent samples (i.e., solvent marked for disposal or reclamation) contains adequate inhibitors. In the case of the MC samples, the inhibitor concentration was approximately that of new solvent and the solvent appeared to the naked eye to be "water white." Furthermore, monitoring the additive concentration in PERC samples showed the inhibitor concentration in the spent solvent to be higher than that of new solvent. These findings suggest inadequate monitoring of solvent inhibitor levels.

The comparison of AAV-versus-usage time curves with those of inhibitor concentration versus time showed good agreement between acid acceptance and acid acceptor concentration. However, acid acceptance as a criterion was inadequate for determining the concentration of antioxidants or metal stabilizers. At present, metal stabilizer levels are monitored by the aluminum "scratch" test. However, this test shows only if the stabilizer levels are adequate or inadequate and cannot give any indication of the metal stabilizer concentration. A field test for determining the levels of these additives would enable the user to more closely monitor a solvent's stabilizer concentration solvent and would consequently allow safer, more efficient operation. The two metal stabilizer tests discussed earlier show promise of being developed into simple field tests.

Solvent Reclamation Studies

These studies showed that distillation is extremely well suited for recovering these chlorinated solvents. Using this method, oil and other contaminants can be removed from the solvent with good recovery of the inhibitors. Activated carbon adsorption was found to be less well suited for reclaiming used solvent. It seemed to selectively adsorb certain additives, severely depleting them (Tables 18 through 20).

Inhibitor Kinetic Studies

Batch kinetic studies showed the reactions between common acid acceptors (butylene oxide, cyclohexene oxide, and epichlorohydrin) and HCl in the solvents to be second-order overall. The rate expressions for the reactions of these compounds with HCl in PERC are as follows:

$$-r_1 = k_1 \text{ [Butylene Oxide][HCl], mol/L} \cdot \text{sec} \qquad \text{[Eq 19]}$$

$$-r_2 = k_2 \text{ [Cyclohexene Oxide][HCl], mol/L} \cdot \text{sec} \qquad \text{[Eq 20]}$$

$$-r_3 = k_3 \text{ [Epichlorohydrin][HCl], mol/L} \cdot \text{sec} \qquad \text{[Eq 21]}$$

Also, the temperature dependencies of the rate constants can be expressed as follows:

Butylene Oxide:

$$k_1 = 9.1 \times 10^6 \exp(-4.5 \times 10^3/T), \text{ L/mol} \cdot \text{sec} \qquad \text{[Eq 22]}$$

Cyclohexene Oxide:

$$k_2 = 1.8 \times 10^5 \exp(-2.7 \times 10^3/T), \text{ L/mol} \cdot \text{sec} \qquad \text{[Eq 23]}$$

Epichlorohydrin:

$$k_3 = 2.8 \times 10^6 \exp(-4.8 \times 10^3/T), \text{ L/mol} \cdot \text{sec} \qquad \text{[Eq 24]}$$

The batch reaction kinetic studies showed the respective reactions between (1) dioxane and $AlCl_3$ and (2) dioxolane and $AlCl_3$ to be second-order overall. The rate expressions can be written as:

$$-r_4 = k_4 \text{ [Dioxane][AlCl}_3\text{], mol/L} \cdot \text{sec} \qquad \text{[Eq 25]}$$

$$-r_5 = k_5 \text{ [Dioxolane][AlCl}_3\text{], mol/L} \cdot \text{sec} \qquad \text{[Eq 26]}$$

The rate constants for these reactions can be expressed as a function of temperature:

Dioxane:

$$k_4 = 10 \exp(-2.1 \times 10^3/T), \text{ L/mol} \cdot \text{sec} \qquad \text{[Eq 27]}$$

Dioxolane:

$$k_5 = 8.2 \times 10^3 \exp(-3.5 \times 10^3/T), \text{ L/mol} \cdot \text{sec} \qquad \text{[Eq 28]}$$

There was one interesting finding from the study of the nitromethane/$AlCl_3$ reaction. The results showed that nitromethane and $AlCl_3$ in PERC reacted very quickly (90 percent in 1 min). However, Van Gemert found that nitromethane and $AlCl_3$ reacted very slowly in MC (25 percent in 200 min). He also observed solvent degradation.[72] The differences in reaction rates can possibly be explained in terms of solvent interaction (or in the case of PERC, lack of interaction). The MC reacts with $AlCl_3$, thus slowing the reaction between nitromethane and $AlCl_3$.

Determination of Solvent Life

Studies performed at Auburn University showed there to be very little change in these cleaning solvents other than contamination with oil/soils and depletion of inhibitors. Reclaimed solvents appeared to be nearly identical to new solvent except in inhibitor concentration. A program of close monitoring of inhibitor levels during operation combined with the use of an effective reclamation method for used solvent (i.e., distillation) and periodic replacement of the solvent inhibitors lost during operation should maximize the useful lifetime of these solvents and also reduce the production of hazardous wastes.

---

[72]B. Van Gemert.

5. Assessment of Evaporative Losses and Vapor Recovery Techniques

Overview

For this part of the study, USA-CERL evaluated vapor recovery techniques reported in the literature. Before discussing evaporative losses of solvent and methods of recovery, an introduction to degreaser design and operation will be helpful. A vapor degreaser is basically a tank with a heater at its bottom to boil the solvent (Figure 68). Articles to be cleaned are suspended in the vapor zone above the boiling liquid. Solvent vapors condense on the relatively cool parts to dissolve contaminants and provide a continuous rinse in clean solvent. The condensed solvent and contaminants drain from the part and return to the boiling liquid reservoir.[73]

Vapor cleaning often is combined with mechanical action such as liquid immersion or spraying the part with liquid solvent under the vapor zone. The part is held in the vapor zone until it reaches the vapor temperature, which causes condensation to stop. The work is then removed from the vapor degreaser.

Figure 68 shows the sections of a simple vapor degreaser. The lower solvent heating chamber is fitted with heating coils that use either steam, electricity, or gas. The vapors displace air within the degreasing compartment, thus providing a cleaning vapor zone. The height of the vapor zone is controlled by condenser coils located in the sidewalls of the degreaser's upper section. The heat exchange fluid is generally water, which condenses the solvent vapors.

The freeboard is the distance from the top of the vapor zone to the top of the tank. Solvent can be lost into the environment by three mechanisms: (1) diffusion through the freeboard zone, (2) entrainment with the work, which is termed "dragout," and (3) exhaust. By monitoring these solvent vapor losses or emissions, an installation would be following regulations imposed by EPA and OSHA and could reduce new solvent makeup costs substantially.

Emissions

For this discussion, emissions are defined as vapors that diffuse and convect from a degreaser. Emissions from open-top degreasers can occur as described above--by diffusion, dragout, and exhaust. The three main emission categories are shown in Figure 69.

Diffusion is the loss of solvent vapor from a vapor zone. The air/solvent interface at the top of the vapor zone can be disturbed by drafts or movement of the work into and out of the degreaser. When this happens, solvent vapors diffuse out of the freeboard region and into the environment. Another souce of solvent loss is the convection of warm, solvent-rich air from the freeboard due to drafts.

In a smoothly operating degreaser, the solvent vapor generation rate is very close to the solvent condensation rate on the work. If excessive vapor generation occurs, the condenser will not be able to handle the high cooling requirement, thus resulting in

---

[73]T. J. Kearney and C. E. Kircher, April 1960 and May 1960; R. Monahan; R. L. Marinello; *ASTM Handbook of Vapor Degreasing.*

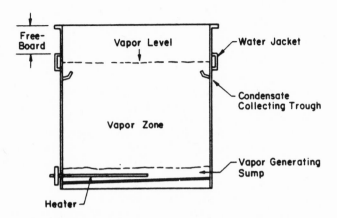

Figure 68. Basic vapor degreaser. (Source: T. J. Kearney and C. E. Kircher, "How To Get the Most From Solvent-Vapor Degreasing--Part I," *Metal Progress* [April 1960], pp 87-92. Used with permission.)

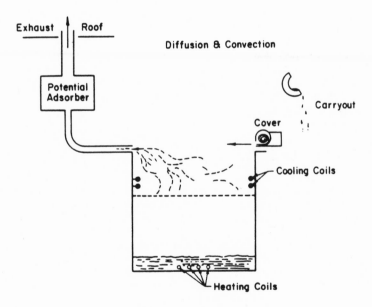

Figure 69. Open-top vapor degreaser emission points. (Source: *Source Assessment: Solvent Evaporation-Degreasing Operations*, EPA-600/2-79-019f [U. S. Environmental Protection Agency, August 1979], p 133.)

solvent vapor loss. On the other hand, a low vapor generation rate causes the vapor zone to drop and draws air into the degreaser. This air and vapor mixture is highly susceptible to drafts.[74]

Dragout emissions are the liquid and vapor of the solvent entrained on the clean parts as they are removed from the degreaser. The parts may contain isolated pockets of entrained liquid or vapor even after drying in the degreaser. Removing the parts disturbs the air/vapor interface, causing the convective dragout of solvent vapors. The vapor level can also be disturbed when an oversized workbasket is used. Besides adding to the energy requirement for heating, such baskets act as pistons, disturbing the vapor level and causing a loss of solvent.

Vapor degreasers may have an open-top or a top-closed conveyorized design. Exhaust systems are installed on large, open-top degreasers for operator safety and plant protection. Some systems provide a carbon adsorption system to collect emission for possible reuse. An efficient carbon adsorption system can completely eliminate exhaust emissions.[75]

Conveyorized degreasers have the same emission categories as open-top degreasers. However, the diffusive and convective vapor losses are less because of the closed top. Figure 70 shows the emission points of an enclosed conveyorized degreaser. The carryout or dragout emission is the most significant category of solvent loss by emission.

Some examples of open-top degreasers are:[76]

- Vapor-distillate spray machine
- Vapor-spray-vapor degreaser
- Liquid-vapor degreaser
- Two-chamber immersion degreaser
- Multiple immersion degreaser
- Ultrasonic degreaser
- Conventional degreaser.

Conveyorized vapor degreasers differ from their open-top counterparts only in the mechanism of material handling. Open-top degreasers employ hand-held baskets or overhead cranes, whereas conveyorized degreasers have little or no manual parts handling. In addition, most conveyorized degreasers are covered at the top.

Examples of conveyorized degreasers are:[77]

- Cross-rod degreaser
- Monorail vapor degreaser
- Vibra degreaser
- Ferris wheel degreaser
- Belt degreaser

---

[74] R. Monahan; R. L. Marinello; *ASTM Handbook of Vapor Degreasing*.
[75] USEPA, August 1979.
[76] R. W. Bee and K. E. Kawaoka; T. J. Kearney and C. E. Kircher, April 1960 and May 1960; R. Monahan; R. L. Marinello; *ASTM Handbook of Vapor Degreasing*.
[77] USEPA, August 1979; T. J. Kearney and C. E. Kircher, April 1960 and May 1960; R. Monahan; R. L. Marinello; *ASTM Handbook of Vapor Degreasing*.

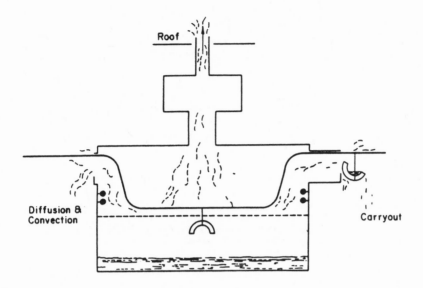

Figure 70. Conveyorized degreaser emission points. (Source: *Source Assessment: Solvent Evaporation-Degreasing Operations*, EPA-600/2-79-019f [U. S. Environmental Protection Agency, August 1979], p 133).

- Strip degreaser
- Circuit board degreaser.

Further information on the two types of degreaser systems are available in the literature (see footnote 77).

Quantitative Emission Determination

Emission for specific degreasing operations is determined by tracking the total amount of solvent supplied to the operation and subtracting from it the amount accountable through degreaser waste solvent activities or, in other words, by performing a solvent mass balance around the degreaser.

The ratio of the amount of solvent unaccounted for to the amount of solvent supplied is termed the "emission factor." Hooghem, et al. found that, on average, the emission factors for open-top and conveyorized vapor degreasing in United States are 775 and 850 g/kg solvent consumed, respectively.[78]   Tables 21 and 22 list characteristics of emissions from typical open-top and conveyorized vapor degreasing operations.

---

[78]USEPA, August 1979.

Table 21

Characteristics of Emissions From Representative
Open-Top Vapor Degreasing Operations*

Solvent	Average Degreaser Size (kg Solvent Consumed/yr)	Average Height (m)	Frequency of Operation (%)	Emission Rate (g/sec)**
Fluorocarbons	3,806	10.6	65	0.1439
PERC	10,070	10.7	78	0.3173
TCE	7,165	12.0	78	0.2257
MC	16,394	14.1	96	0.4197

*Source: Source Assessment: Solvent Evaporation-Degreasing Operations, EPA-600/2-79-019f (U. S. Environmental Protection Agency, August 1979), p 133.
**Emission rate = $\left[\dfrac{\text{average solvent consumption/yr x emission factor}}{\text{frequency of operations x no. of sec/yr}}\right]$

Table 22

Characteristics of Emissions From Representative
Conveyorized Vapor Degreasing Operations*

Solvent	Average Degreaser Size (kg Solvent Consumed/yr)	Average Height (m)	Frequency of Operation (%)	Emission Rate (g/sec)**
Fluorocarbons	9,403	10.6	65	0.3899
PERC	24,883	10.7	78	0.8598
TCE	17,780	12.0	78	0.6144
MC	40,468	14.1	96	1.1362

*Source: Source Assessment: Solvent Evaporation-Degreasing Operations, EPA-600/2-79-019f (U. S. Environmental Protection Agency, August 1979), p 133.

**Emission rate = $\left[\dfrac{\text{average solvent consumption/yr x emission factor}}{\text{frequency of operations x no. of sec/yr}}\right]$

Emission Control

Several preventive measures can significantly reduce emissions from vapor degreasers. They are:

1. High/increased freeboard
2. Refrigerated chillers placed above condensers
3. Enclosed or covered operations
4. Secondary recovery by carbon adsorption
5. Safety switches.

Increased Freeboard. The freeboard reduces drafts near the air-solvent interface. Open-top degreasers should have a freeboard-to-width ratio of 0.75 or greater. Studies indicate that increasing freeboard-to-width ratio from 0.5 to 1.0 may greatly reduce emissions of open-top degreasers. For an enclosed conveyorized degreaser, the freeboard-to-width ratio must be at least 0.5, with a minimum height of 36 in.[79]

Refrigerated Chillers. The vapor created within a degreaser is condensed by condenser coils located above a vapor zone at the beginning of freeboard zone. This cooling system can be augmented by refrigerated freeboard chillers placed just above the condenser coils. Although at first glance the chillers may give the appearance of a secondary cooling system to support the condenser, the actual purpose is much different. The condenser establishes the upper unit of the vapor zone, whereas the chillers slow the diffusive emission from the vapor zone into the freeboard and subsequently the atmosphere.

The chiller should be capable of removing at least 100 Btu/hr/ft of perimeter coil. The chilling of air by the chiller creates a cold air blanket. This air blanket establishes a sharp temperature gradient which reduces the mixing of air and solvent vapor by narrowing the air/vapor mixing zone. Also, the chilled blanket decreases convection of warm, solvent-rich air.

Another variation of a refrigerated chiller is the refrigerated condenser coil. In this design, the primary condenser and the freeboard chiller are replaced by refrigerated condenser coils. Coolant in the condenser coil creates a blanket of cold air above the air/vapor interface.

The refrigerated chiller may decrease emissions by about 40 percent.[80] The coolant in a chiller can be operated at above-freezing as well as below-freezing temperatures. However, the effectiveness of above-freezing temperature chillers are not as well established as that of below-freezing operations.

Enclosed or Covered Degreasers. Installing a hood or cover is the simplest way of emission control for open-top degreasers. The covers are generally designed to open and close in a horizontal motion so that the air/vapor interface is not disturbed. Covers are available in several designs, including roll-type plastic covers, canvas curtains, and guillotine covers.

Automatic covers are designed to open for the time it takes work to be introduced into and out of a degreaser. Although conveyorized degreasers include a cover, additional emission prevention covers can be added. Covers for the entrance and exit of conveyorized degreasers impedes drafts into the degreasers. This type of cover especially helps in reducing evaporative losses after shutdown of the degreaser because the hot solvent is cooled by evaporation.

A cover on an open-top vapor degreaser has been shown to reduce total emissions by 20 percent to 40 percent;[81] for conveyorized degreasers, approximately 18 percent of the total emissions are due to evaporation during downtime, all of which can be eliminated by using a cover. However, it must be stressed that covers can help retard

[79]T. J. Kearney and C. E. Kircher, April 1960 and May 1960; R. Monahan; R. L. Marinello; ASTM Handbook of Vapor Degreasing.
[80]USEPA, August 1979.
[81]USEPA, August 1979.

emission only when solvent evaporation accounts for the major portion of total emissions. The cover cannot help reduce emissions if other types of emission are dominant (e.g., dragout).

Secondary Recovery by Carbon Adsorption. As defined here, adsorption is the process of removing gas molecules from a stream through contact with a solid. The solid stationary phase is the adsorbent and the adsorbed gas is the adsorbate. By using activated carbon as adsorbent, chlorinated solvent vapors can be adsorbed and recovered for reuse.[82]

The adsorbents are highly porous materials with a large surface area (i.e., large surface-to-volume ratios) and specific affinity for individual adsorbates. Activated carbon can adsorb more than 90 percent of many organic vapors from air at ambient temperature. The quantity of organic vapor adsorbed by activated carbon depends on the type of adsorbate as well as its temperature and concentration. The amount of vapor adsorbed increases with increasing vapor concentration but decreases with increasing vapor temperature.

The effect of reclamation via carbon adsorption on solvent inhibitors was discussed in Chapter 4. Carbon adsorption systems for cleaning solvents can achieve only 40 to 65 percent reduction of the total emission. This is due to the fact that the major loss areas are usually dragout on parts, leaks, and spills.[83]   Improved ventilation design may increase an absorber's overall emission control efficiency.

Tests conducted on carbon adsorption systems over an open-top vapor degreaser revealed a 60 percent emission reduction.[84]   However, many adsorption systems yield less than 40 percent emission reduction, either due to poor inlet collection efficiency or improper maintenance of the carbon adsorber. The inlet collection efficiency is the percentage of solvent vapors emitted by the degreaser that are captured by the inlet piping of the carbon adsorption system.

Safety Switches. These switches are preventive devices installed on vapor degreasers and are activated only during a malfunction. There are five main types of safety switches: (1) safety vapor thermostat, (2) condenser water flow switch and thermostat, (3) sump thermostat, (4) solvent level control, and (5) spray safety switch.[85]   Of these switches, the safety vapor thermostat is the most important. This device detects the solvent vapor zone when it rises above the condenser coils and turns off the heat.

Liquid Absorption. TCE vapors in air can be reduced by absorption in mineral oil. However, this operation may release mineral oil into the environment (120 ppm at 30 °C).[86]   The net effect is to replace the emission of one hydrocarbon with another. Unless very toxic or highly valuable vapors are involved, this method of emission control is not practical.

---

[82]USEPA, August 1979; B. L. Brady, Jr., *Study of Chlorinated Solvents*, Master's Thesis (Auburn University, 1987).
[83]USEPA, August 1979.
[84]USEPA, August 1979.
[85]USEPA, August 1979.
[86]USEPA, August 1979.

6. Metal Preparation and Precision Cleaning Solvents

The same physicochemical tests described in Chapter 3 were performed on two metal preparation and precision cleaning solvents--IPA and freon-113. The solvents were obtained from the gyro and avionic shops of Robins AFB, GA. Some of the process data for these two solvents are listed in Table 23. The two chemicals were used only as spray-wash solvents. The samples obtained were those of new, recycled, and spent wash solvents.

Tests performed on the samples were:

- KBV
- Viscosity
- Specific gravity
- Refractive index
- Electrical conductivity
- Visible absorbence spectroscopy.

The results are described below and are summarized in Tables 24 and 25.

1,1,2-Trichloro-1,2,2-trifluoroethane (Freon-113)

Kauri-Butanol Value (KBV)

Figure 71 shows the KBV profile of the freon-113 samples. The spent solvent showed a slightly higher KBV over the new and recycled solvents by 6 and 3 percent, respectively. This result was probably due to contamination during cleaning. The higher KBV of the recycled solvent compared with the new solvent indicated that traces of the contaminant carry over into the distillate. Since KBV indicates solvent power, the recycled solvent still had the cleaning potency of a new solvent according to these results.

Table 23

Process Data for Precision Cleaning Fluids*

Solvent	Solvency for Metal-working Soils	Toxicity (ppm)	Flash Point (°F)	Evaporation Rate	Water Solubility (% of Weight)	Boiling Point Range (°F)	Weight (lb/gal)
Freon-113	Good	1000	None	1.7 ($CCl_4=1$)	<0.1	117	13.16
IPA	Poor	400	55	0.07 ($CCl_4=1$)	Infinite	179 to 181	6.55

*Source: R. L. Marinello, "Metal Cleaning Solvents," *Plant Engineering* (30 October 1980), pp 52-57. Used with permission.

Table 24

Experimental Physical Properties of Freon-113

Property	New	Spent	Recycled
Kauri-butanol value	30.8000	32.5000	31.5000
Viscosity (25°C), cp	0.6600	0.6300	0.6700
Specific gravity (25°C/25°C)	1.5690	1.5600	1.5700
Refractive index (20.5°C)	1.3586	1.3591	1.3590
Electrical conductivity (26°C), nanomho/cm	25.9000	24.7000	25.9000

Table 25

Experimental Physical Properties of IPA

Property	New	Spent	Recycled
Kauri-butanol value[*]	----	----	----
Viscosity (30°C), cp	1.7600	1.6300	1.6800
Specific gravity (25°C/25°C)	0.7830	0.8240	0.8030
Refractive index (20.5°C)	1.3762	1.3776	1.3781
Electrical conductivity (26°C), micromho/cm	0.2200	6.1700	1.2600

[*]No KBV endpoint was observed.

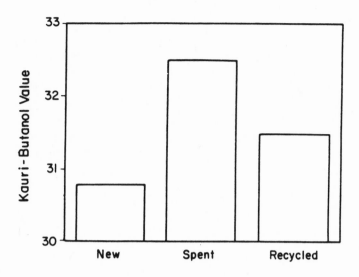

Figure 71. Kauri-butanol value of freon-113 samples.

Viscosity

The viscosity of freon-113 samples at 25 °C (Figure 72), indicates that the spent solvent decreased in viscosity by about 5 percent with respect to the new solvent. This finding contradicts the observations with Stoddard and chlorinated solvents, for which the viscosity increased with contamination. The recycled and new freon-113 viscosities were essentially the same. The viscosity data reported in the literature for freon-113 at various temperatures are listed in Table 26.

Specific Gravity

Specific gravities of the three freon-113 samples at 25 °C are graphed in Figure 73. The spent solvent specific gravity increased by about 5 percent over that of new solvent. The recycled solvent specific gravity was about 2 percent higher relative to that of new solvent. Published data on saturated freon-113 density are listed in Table 27.

Refractive Index

Figure 74 shows the variation in refractive index for the freon-113 samples at 20.5 °C. The spent solvent showed an increase of 0.027 percent in refractive index compared with new solvent. (The earlier experiments with chlorinated solvents indicated that refractive index was neither a very sensitive nor consistent test method.)

Electrical Conductivity

The electrical conductivity profile of freon-113 at 23 °C is shown in Figure 75. The spent solvent had a 6 percent lower conductivity than new solvent. The recycled solvent conductivity was essentially the same as new solvent. Experience with other cleaning solvents has indicated that contaminants, in small concentrations, decrease conductivity.

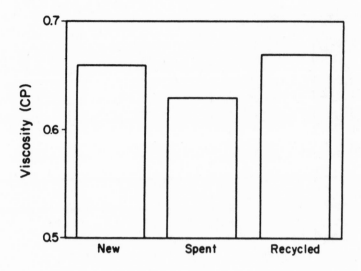

Figure 72. Viscosity of freon–113 samples.

Table 26

Viscosity of Freon–113

Temperature (°C)	10	20	30	40
Viscosity (cp)	0.82	0.67	0.62	0.55

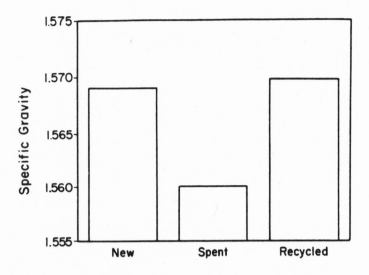

Figure 73. Specific gravity of freon-113 samples.

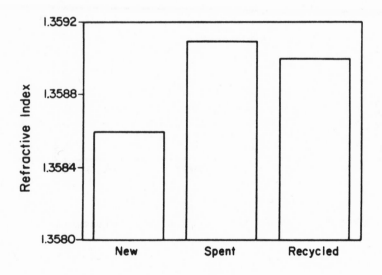

Figure 74. Refractive index of freon-113 samples.

Table 27

Saturated Freon-113 Density

Temperature (°F)	0	40	80	100	150	200
Pressure (psia)	0.84	2.66	6.90	10.48	25.93	54.66
Density (lb/cu ft)	103.5	100.6	97.5	95.8	91.4	86.7

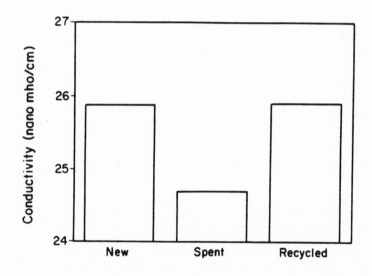

Figure 75. Electrical conductivity of freon-113 samples.

Visible Absorbence

The solvents were measured at four wavelengths (400, 450, 500, and 600 nm). Generally, lowering the wavelength increased the sensitivity of this test method as can be seen in Figure 76. Spent solvent can easily be distinguished from new and recycled solvents by visible absorbence.

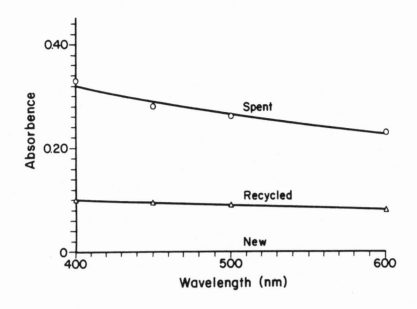

Figure 76. Visible absorbence of freon-113 samples.

Isopropanol (IPA)

Kauri-Butanol Value (KBV)

No KBV endpoint was obtained for IPA.

Viscosity

IPA viscosity was approximately three times that of freon-113 at 25 °C. The viscosity of IPA samples at 25 °C indicates that the spent solvent decreased in viscosity by about 7.3 percent compared with new solvent (Figure 77). The recycled solvent viscosity was about 4.8 percent less than the new solvent viscosity. Table 28 lists viscosity data on IPA as reported in the literature.[87]

Specific Gravity

The specific gravity of IPA at 25 °C is plotted in Figure 78. The spent solvent specific gravity increased by about 5.2 percent over that of new solvent. The recycled solvent had a 2.5 percent increase in specific gravity over new solvent. The literature data on IPA specific gravity are listed in Table 29.

---

[87] *Handbook of Chemistry and Physics; Chemical Engineers' Handbook,* R. H. Perry and C. H. Chilton (Eds.), 5th ed. (McGraw-Hill, 1973).

Refractive Index

The variation in refractive index (21.5 °C) of IPA samples is shown in Figure 79. The spent and recycled solvent had higher refractive indices than the new solvent (by 0.14 and 0.10 percent, respectively).

Electrical Conductivity

Figure 80 shows the electrical conductivity profile of IPA at 26 °C. The spent solvent had a 26.7 times higher conductivity than the new solvent. The recycled solvent was about 4.7 times more conductive than new solvent. This finding was an anomaly because most degreaser contaminants, in small concentrations, decrease the conductivity of chlorinated solvents. The electrolytic conductivity of IPA, as reported in the literature, is 3.5 micromho/cm at 25 °C.[88]

Visible Absorbence

Visible absorbence was measured at four wavelengths (400, 450, 500, and 600 nm). As in the case of the other solvents studied, lowering the wavelength increased the sensitivity of this test method (Figure 81). The spent solvent could easily be distinguished from the new and recycled solvents using visible absorbence.

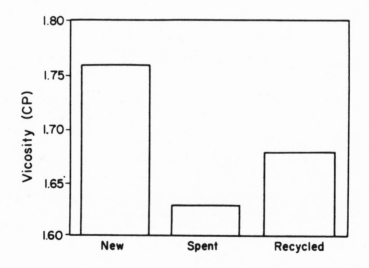

Figure 77. Viscosity of isopropyl alcohol samples.

[88] *Handbook of Chemistry and Physics.*

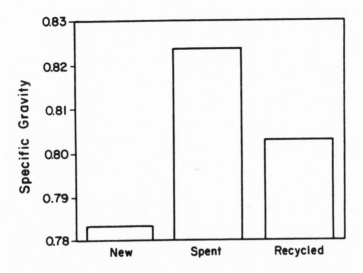

Figure 78. Specific gravity of isopropyl alcohol samples.

Table 28

Viscosity of IPA

Temperature (°C)	10	20	30	40
Viscosity (cp)	3.02	2.32	1.76	1.37

Table 29

Specific Gravity of IPA

Temperature (°C)	0	15	25	30
Specific Gravity	0.802	0.789	0.785	0.777

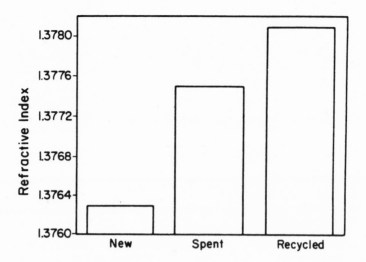

Figure 79. Refractive index of isopropyl alcohol samples.

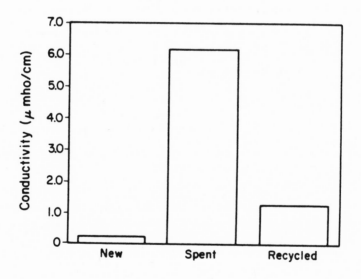

Figure 80. Electrical conductivity of isopropyl alcohol samples.

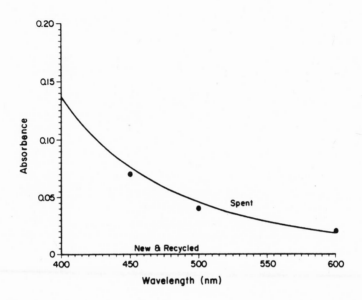

Figure 81. Visible absorbence of isopropyl alcohol samples.

7. Conclusions and Recommendations

This study has evaluated simple tests to be used as criteria in determining the condition of cleaning solvents at U.S. Army installations. Reliable tests that could be performed easily onsite would ensure that a solvent's useful life is maximized, reducing the costs of new solvent purchase and waste disposal. This second phase of the work has focused on halogenated compound solvents used in vapor degreasing and metal/precision cleaning operations.

The physicochemical tests selected for evaluation were those judged most sensitive to solvent contamination and most likely to be easy to perform in the field with reliable results. To assess these tests, solvent samples were obtained from several installations; the samples represented different stages of solvent life based on the length of time in use. Visible absorbence spectrometry was found to be the most reliably measured property, with AAV also yielding good results. The most accurate way to assess solvent quality would be to use visible absorbence combined with at least one other test.

Additional testing is required for chlorinated solvents to determine inhibitor concentrations. Inhibitors are added to these solvents by the manufacturer and are critical to safe, efficient use. Therefore, a chlorinated solvent can remain useful only as long as the inhibitor level is adequate. For the solvents studied in this phase of the project, the major inhibitors were identified using GC-MS. Inhibitor concentrations were monitored with usage time for several batches of solvent. Batch kinetic studies showed that reactions between the acid acceptors and hydrochloric acid, and between the metal stabilizers and aluminum chloride were each second-order overall when tetrachloroethylene was used as solvent. As part of this phase, a simple test was devised to determine metal stabilizer concentration in a solvent.

Reclamation studies of spent chlorinated solvents indicated that batch distillation is well suited for recovery of these solvents, with inhibitors remaining at a safe level. The other method evaluated--activated carbon absorption--was unsatisfactory. This method selectively adsorbed certain inhibitors almost completely.

In addition to maximizing solvent life, other steps feasible for cost-effective solvent management are reduction of evaporative losses and vapor recovery. A substantial amount of solvent is lost through evaporation and dragout, which has implications not only for replacement costs, but also for environmental concerns. Several effective methods are available for minimizing emissions from a degreaser and for capturing and reclaiming the solvent.

Tests for metal/precision cleaners were evaluated separately. Except for refractive index, all other test methods studied for freon-113 had sensitivities of 5 percent or higher. In the case of IPA, electrical conductivity was by far the most sensitive method, although all other tests (excluding refractive index) showed reasonable sensitivity.

This study has shown that a careful program of solvent monitoring to check inhibitor levels during operation, combined with an effective reclamation method and periodic replacement of inhibitors, should maximize the useful lifetime of halogenated solvents and reduce the amount of hazardous waste produced. Specific recommendations are summarized as follows:

1. Visible absorbence and AAV should be used as primary criteria to identify spent solvents. Use of one or more additional tests will increase reliability.

2. Installations can develop "cutoff" values for the tests based on measurement or experience with cleaning jobs onsite. These threshold values will simplify the measuring process and help field personnel make intelligent decisions with regard to changing solvents.

3. Batch distillation should be considered as an effective way to reclaim spent solvent. The batch distillation process will be most feasible for units that can be used to reclaim multiple solvents. Inhibitor levels must be monitored closely using AAV as the primary criterion; if the concentration falls below a safe level, makeup inhibitor must be added.

4. For effective control of emissions, open-top degreasers should have a freeboard-to-width ratio of 0.75 or greater (48 in. maximum). Degreasers should be covered when not in use. Conveyorized degreasers should have a freeboard-to-side ratio of at least 0.5 with a maximum height of 36 in. (Note: minimum freeboard height may be stipulated by existing EPA or state air pollution regulations.)

5. A freeboard chiller located above the condenser coils and with a duty of at least 100 Btu/hr-ft of perimeter coil is highly desirable.

6. The workbasket should not be oversized for two reasons: to avoid a waste of energy for heating the basket and to prevent a piston effect that will disturb the vapor level and cause a loss of solvent.

7. Emissions from an open top can be adsorbed for reclamation by a standard carbon adsorption unit.

References

Abbe–56 Refractometer Manual (Bausch and Lomb Optical Co.).

Annual Book of ASTM Standards (1987).

Archer, W. L., *Ind. Eng. Chem., Prod. Res. Dev.*, Vol 21 (1982).

Archer, W. L. and E. L. Simpson, *Ind. Eng. Chem., Prod. Res. Dev.* Vol 16, No. 2 (1977).

Archer, W. L., E. L. Simpson, and R. R. Gerard, U. S. Patent 4,018,837 (1977).

Ashworth, M. R. F., *Titrimetric Organic Analysis Part I: Direct Methods* (Interscience Publishers, 1964).

ASTM Handbook of Vapor Degreasing, ASTM Special Technical Publication No. 310 (American Society for Testing and Materials [ASTM], April 1962).

Bauer, H. H., G. D. Christian, and J. E. O'Reilly, *Instrumental Analysis* (Allyn and Bacon, 1979).

Beckers, N. L., U. S. Patent 3,796,755 (1974).

Beckers, N. L., and E. A. Rowe, U. S. Patent 3,935,287 (1976).

Beckers, N. L., and E. A. Rowe, U. S. Patent 3,957,893 (1976).

Bee, R. W., and K. E. Kawaoka, *Evaluation of Disposal Concepts for Used Solvents at DOD Bases*, TOR-0083(3786)-01 (The Aerospace Corporation, February 1983).

Borror, J. A., and E. A. Rowe, Jr., U. S. Patent 4,293,433 (1981).

Brady, B. L., Jr., *Study of Chlorinated Solvents*, Master's Thesis (Auburn University, 1987).

Bunge, A.L., *Minimization of Waste Solvent: Factors Controlling the Time Between Solvent Changes*, CERL Contract No. DACA 88-83-C-0012 (Colorado School of Mines, September 1984).

Castrantas, H. M., R. E. Keay, and D. G. MacKellar, U.S. Patent 3,677,955 (1972).

Chemical Engineers' Handbook, R. H. Perry, and C. H. Chilton (Eds.), 5th ed. (McGraw-Hill, 1973).

Chlorinated Solvent Information, Fact Sheet No. 5 (Dow Chemical Co.).

Cormany, C. L., U. S. Patent 4,065,323 (1977).

Culver, M. J., and H. R. Stopper, U. S. Publ. Pat. Appl. B US 370,309 (1976).

Esposito, G. G., *Solvency Rating of Petroleum Solvents by Reverse Thin-Layer Chromatography*, AD-753336 (Coating and Chemical Laboratory, Aberdeen Proving Ground, 1972).

Gutsche, C. D., and D. J. Pasto, *Fundamentals of Organic Chemistry* (Prentice-Hall, 1975).

Handbook of Chemistry and Physics, 57th ed. (CRC Press, 1976).

International Fabricare Institute (IFI) Bulletin, T-447 (1969).

Ishibe, N., and J. K. Harden, U. S. Patent 4,368,338 (1983).

Johnson, K., *Drycleaning and Degreasing Chemicals and Processes* (Noyes Data Corp., 1973).

Kearney, T. J., and C. E. Kircher, "How To Get the Most From Solvent-Vapor Degreasing--Part I," *Metal Progress* (April 1960), pp 87-92.

Kearney, T. J., and C. E. Kircher, "How To Get the Most From Solvent-Vapor Degreasing--Part II," *Metal Progress* (May 1960), p 93.

Lee, H. J., I. H. Custis, and W. C. Hallow, *A Pollution Abatement Concept, Reclamation of Naval Air Rework Facilities Waste Solvent, Phase I* (Naval Air Development Center, April 1978).

Levenspiel, O., *Chemical Reaction Engineering*, 2nd ed. (John Wiley and Sons, 1972).

Lundberg, W. O., *Symposium on Food: Lipids and Their Oxidation*, H. W. Schultz, et al. (Eds.) (AVI Publishing Co., 1962).

Manner, J. A., U. S. Patent 3,532,761 (1970).

Manner, J. A., U. S. Patent 4,026,956 (1977).

Marinello, R. L., "Metal Cleaning Solvents," *Plant Engineering* (October 30, 1980), pp 52-57.

McDonald, L. S., U. S. Patent 3,565,811 (1971).

Mellan, I., *Industrial Solvents* (Reinhold, 1950).

Mettler/Paar DMA 35 Density Meter (Mettler Instrument Corp., 1986).

Monahan, R., "Vapor Degreasing With Chlorinated Solvents," *Metal Finishing* (November 1977), pp 26-31.

National Institute of Drycleaning (NID) Bulletin Service, T-413 (1965).

Niven, W. W., *Fundamentals of Detergency* (Reinhold, 1950).

Peoples, L., U. S. Patent 3,746,648 (1973).

Pereira, A. S., *Composition and Stability of Poultry Fats*, Ph.D. Thesis (Purdue University, 1975).

Phillips, E. R., *Drycleaning* (National Institute of Drycleaning, 1961).

Rains, J. H., U. S. Patent 3,629,128 (1971).

Richtzenhain, H., and R. Stephan, U. S. Patent 3,787,509 (1974).

Richtzenhain, H., and R. Stephan, U. S. Patent 3,959,397 (1976).

Shugar, G. J., et al., *Chemical Technician's Ready Reference Handbook*, 2nd ed. (McGraw-Hill, 1981).

Spencer, D. R. and W. L. Archer, U. S. Patent 4,115,461 (1978).

Stenhagen, E., et al., *Registry of Mass Spectral Data*, Vol 1 (John Wiley and Sons, 1974).

Tipping, J. W., Brit. GB 1,276,783 (1972).

U.S. Environmental Protection Agency (USEPA), *Source Assessment: Reclaiming of Waste Solvents, State of the Art*, EPA-600/2-78-004f (Industrial Environmental Research Laboratory, April 1978).

USEPA, *Source Assessment: Solvent Evaporation-Degreasing Operations*, EPA-600/2-79-019f (Industrial Environmental Research Laboratory, August 1979).

Van Gemert, B., Brit. UK Pat. Appl. GB 2,027,697 (1980).

Van Gemert, B., *Ind. Eng. Chem., Prod. Res. Dev.*, Vol 21 (1982).

Willard, H. H., L. L. Merritt, Jr., and J. A. Dean, *Instrumental Methods of Analysis*, 5th ed. (D. Van Nostrand, 1974).

Willard, H. H., L. L. Merritt, Jr., J. A. Dean, and F. A. Settle, Jr., *Instrumental Methods of Analysis*, 6th ed. (Wadsworth Publishing, 1981).

Wingrove, A. S., and R. L. Caret, *Organic Chemistry* (Harper and Row, 1981).

Abbreviations

AAV	acid acceptance value
AFB	Air Force Base
ASTM	American Society for Testing and Materials
BP	boiling point
DOD	Department of Defense
ECD	electron capture detector
FEAP	Facilities Engineering Applications Program
FID	flame ionization detector
FPD	flame photometric detector
Freon-113	1,1,2-trichloro-1,2,2-trifluoroethane
GC	gas chromatography
IFI	International Fabricare Institute
IPA	isopropanol, isopropyl alcohol
K-B	Kauri-Butanol solution
KBV	Kauri-Butanol value
MC	1,1,1-trichloroethane, methyl chloroform
MS	mass spectrometer
NARF	Naval Air Rework Facility
NAS	Naval Air Station
NID	National Institute of Drycleaning
OSHA	Occupational Safety and Health Administration
PERC	perchloroethylene, tetrachloroethylene
R_f	response factor
% S	percentage sensitivity
TCD	thermal conductivity detector
TCE	trichloroethylene

TLC	thin-layer chromatography
USA-CERL	U.S. Army Construction Engineering Research Laboratory
USE	Used Solvent Elimination
USEPA	U.S. Environmental Protection Agency